Debate Dynamics: How Controversy Improves Our Beliefs

SYNTHESE LIBRARY

STUDIES IN EPISTEMOLOGY,

LOGIC, METHODOLOGY, AND PHILOSOPHY OF SCIENCE

VOLUME 357

For further volumes:
http://www.springer.com/series/6607

Gregor Betz

Debate Dynamics: How Controversy Improves Our Beliefs

Springer

Gregor Betz
Institute of Philosophy
Karlsruhe Institute of Technology
Kaiserstrasse 12
D-76131 Karlsruhe
Germany

ISBN 978-94-007-4598-8 ISBN 978-94-007-4599-5 (eBook)
DOI 10.1007/978-94-007-4599-5
Springer Dordrecht Heidelberg New York London

Library of Congress Control Number: 2012946202

Printed on acid-free paper

Springer is part of Springer Science+Business Media (www.springer.com)

Acknowledgements

This book would not exist without the generous support I received from the Stiftung Alfried Krupp Kolleg Greifswald. I'm very grateful for the year I was allowed to spend as a fellow in Greifswald, and I'd like to thank Prof. Bärbel Friedrich, Dr. Christian Suhm, and their team personally for the close and cordial collaboration. Moreover, I'm indebted to colleagues and friends who critically assessed parts of the reasoning this book unfolds, specifically to Sebastian Cacean, Moritz Cordes, David Löwenstein, Friedrich Reinmuth, Holm Tetens, and Christian Voigt as well as the participants of two colloquia at the University of Greifswald and at the Freie Universität Berlin. Last but not least, I'd like to thank the anonymous reviewer of the Synthese Library Series for the astute comments and very helpful suggestions.

Contents

Chapter 1
General Introduction

1.1 The Aims of Argumentation

The idea that controversial argumentation improves our beliefs is as old as argumentation theory—the systematic reflection on argumentation—itself and probably even older. But what precisely does this alleged improvement we aim at when engaging in a controversy mean? In which sense does the game of giving and taking reasons, presumably, better our beliefs?

We may discern, by and large, two distinct, fundamental rationales one can pursue in an argumentation. The first one is to overcome dissent and to reach a consensus. The second one consists in rectifying our errors and tracking down the truth.

These two aims can be pursued, and achieved, quite independently of each other. The proponents in a debate may very well reach a consensus position without having acquired correct beliefs, in which case the consensus is a spurious one. Similarly, a proponent might acquire true beliefs, achieving the second rationale, while continuing to disagree with her opponents. But obviously, if all proponents have found the truth, they eo ipso agree.

Proponents who engage in an argumentation don't necessarily strive both for consensus *and* for truth. Some debates, such as, for example, the moral controversy about preimplantation genetic diagnosis or the debate about legalizing voluntary euthanasia, might primarily aim at reaching broad societal agreement, and the proponents might simply not be interested in the additional question whether a consensus position, should it once emerge, is also true (if they judge that question meaningful at all). In other debates, however, finding the truth constitutes the primary, or even the sole rationale. Think of scientific controversies. Peer agreement is not what intrinsically motivated scientists are ultimately striving for. In the end, they aspire to correct answers—say, to the question whether the Earth was completely covered by ice sheets once or what caused ancient civilizations to collapse—and not merely to a consensus. In these debates, agreement is at most of indirect interest, namely insofar as persisting dissent indicates that not all proponents have found the truth yet. Finally, the relative importance of the two

G. Betz, *Debate Dynamics: How Controversy Improves Our Beliefs*,
Synthese Library 357, DOI 10.1007/978-94-007-4599-5_1,

rationales may not only vary from debate to debate, but even within a debate, from proponent to proponent, or from one (historic) phase to another.

In spite of being principally distinct aims, consensus and truth are intricately related. We have already noted that a consensus among proponents represents a necessary condition for all proponents having found the truth.[1] This simple observation—that dissent indicates falsity—raises the more interesting question whether consensus, vice versa, indicates truth as well. Further questions pertain to potential trade-offs between the two rationales. Does, for example, a controversy which effectively generates agreement among proponents enable them to track down truth? Or, does a truth-conducive argumentation tend to obstruct rapprochement of the proponents? We will return to these issues in due course.

We engage in argumentation in order to reach agreement and to find the truth. Putting forward and listening to arguments is no self-sufficient activity, no *l'art pour l'art*. The rationality of a controversial argumentation thus resides in its effectiveness as to realizing its aims. It is an instrumental rationality. We can distinguish, accordingly, an instrumental *consensual* value of argumentation on the one hand, and an instrumental *veritistic* value of argumentation on the other hand. This book amounts to an investigation of both. It studies the consensual and veritistic value of different argumentative practices—practices which diverge in regard to the way proponents put forward arguments, and modify their convictions in the light of newly introduced reasons. One reading of the ensuing inquiry, hence, consists in conceiving it as a contribution to the reliabilistic program of social epistemology (see Goldman 1999).

1.2 An Example of a Controversial Argumentation

Before I set forth the methods and assumptions of the following investigation, we shall consider an example of a controversial debate. This miniature case study illustrates the kind of argumentation to which the formal investigation, unfolded in this book, applies. Moreover, it helps to frame and grasp the rather abstract concepts which are to be introduced henceforth.

The scientific controversy concerning the origin of the so-called Nördlinger Ries—an uncommon, circular depression of the landscape, which circumscribes the town of Nördlingen in Southern Germany and which represents in fact, as we know today, the remnants of an impact crater, testifying to the impact of a meteorite roughly 15 million years ago—will serve as an example.[2] The origin of the Ries, and of its unusual rocks, has been unclear for a long time, and only relatively lately, namely in 1960, did Shoemaker and Chao (1961) succeed in demonstrating that the Ries represents an impact crater, effectively closing a controversy whose beginnings date back to the end of the eighteenth century.

---

[1] As Descartes has already remarked in his *Rules* (Descartes 1984, p. 11).

[2] The following account is based on von Engelhardt (1982) and Kölbl-Ebert (2003).

In search of building materials, the Bavarian engineer C. v. Caspers found, in the 1780s, that specific rocks from the Ries area (referred to as suevite today) are suited for mortar production. Likening these rocks to the trass in the Rhenish area, which possesses similar properties, and whose volcanic origin had only recently been established, Caspers argued:

Hypothesis 1 (Volcanic origin) *Suevite is a volcanic product.*

The volcanic hypothesis has been largely agreed upon in subsequent decades. Geologists who studied the Ries such as Flurl (1805)[3], Schübler (1825), Cotta (1834), and Voith (1835) assented to this theory and, in part, provided additional arguments by drawing further analogies or pointing out that the scattered suevite occurrences may be understood as lava bombs.

Mapping the geology of the Ries area, Schnitzlein and Frikhinger (1848) found that the rocks' sequence in the basin doesn't accord with their normal geological position (i.e., older rocks were situated on top of younger ones). On the basis of this observed stratigraphic disturbance, they argued in favor of an extended and modified volcanic hypothesis.

Hypothesis 2 (Volcanic origin) *Volcanic forces lifted old rocks from deeper depth to the surface, caused multiple eruptions which gave rise to suevite occurrences, and, finally, triggered the subsidence of the basin.*

At the same time, Schafhäutl (1849) proposed a completely different theory, rejecting, in consequence, the volcanic hypothesis.

Hypothesis 3 (Viscous underground) *Vast underground resources of a viscous silicate gel contracted (due to water loss) and sparked off the subsidence; the gel ascended, henceforth, along the resulting fractures at the margins of the depression, and eventually solidified, forming granite-like rocks (including suevite).*

Schafhäutl's main argument was based on a chemical analysis of suevite, which revealed a close resemblance to granite and disclosed, moreover, significant differences between suevite and the Rhenish trass, thereby undermining the argument, originally introduced by Caspers, in favor of the volcanic theory. Given that suevite and granite occurrences are, in addition, locally correlated, Schafhäutl inferred that both stem from one and the same origin, which eventually triggered his inventive hypothesis.

Some 20 years later, Deffner (1870) suggested yet another rival hypothesis.

Hypothesis 4 (Ries glacier) *The Ries basin once hosted a glacier; concentric ice flow in all directions powered a corresponding transport of material.*

Deffner supported his theory on the basis of new evidence pertaining to large amounts of debris, consisting in rocks from different geological periods, outside

[3]For bibliographic references to this as well as to the other original contributions to the Ries controversy mentioned below, compare von Engelhardt (1982).

the Ries basin. Moreover, he had discovered, together with his colleague O. Fraas, signs of lateral transport such as polished surfaces and striations pointing towards the basin's center. A glacier, Deffner thought, was the only mechanism which could account for the vast displacements of rocks, and the characteristic traces which had, at other places, already been attributed to glacial activities.

Fraas, however, well aware of the evidence for massive horizontal displacements, didn't concur in his colleague's glacier hypothesis, questioning, in particular, the ability of a glacier to cause mass transport on the required scale. In addition, he provided a detailed description of the shapes of the suevite bombs, which witness, he argued, to the air resistance in the course of their flight through the atmosphere, and which therefore support the volcanic theory.

Gümbel (1870,91,94), too, accepted the volcanic hypothesis yet argued for a specific modification.

Hypothesis 5 (Volcanic origin) *The suevite occurrences result from a single volcanic ejection.*

He had discovered that all suevite bombs, scattered over the Ries area, display a similar microscopic structure and composition, which suggests that they spring from a single event.

Being apparently still puzzled by the traces of lateral transport such as the polished surfaces and striations, which, allegedly, only pertain to geologically young rocks, Koken (1901,02) argued in favor of a revival of Deffner's glacier hypothesis, which he now understood, however, as a complementary rather than a rival hypothesis to the volcanic theory. By distinguishing different phases of the Ries' evolution, with volcanic activities preceding its later glaciation, Koken tried to reconcile some of the different theories previously proposed.

Branco (1901,03) and Fraas (1901,03,19), however, strictly opposed Koken's modified glacier hypothesis and attempted to refute it on different grounds. They insisted, first, that substantial amounts of debris have been discovered at places far removed from the basin: This documents a mass transport glacier theory cannot account for. Second, the characteristic striations in younger rocks, which supposedly testify to a relatively recent glaciation, are in fact also discernible in much older rocks.

As an alternative to transport by glaciation, Branco proposed the following hypothesis, which still remained, by and large, within the cluster of volcanic theories.

Hypothesis 6 (Ries mountain) *A magma pocket of 25 km diameter created, through expansion, a colossal mountain with steep slopes. As a result of gas release (or other reasons), the mountain eventually disappeared and gave way to today's basin.*

Branco argued that, on account of the Ries mountain's steep slopes, gravitational slides shredded and transported material far beyond today's basin. Yet, realizing that gravitational forces don't yield enough energy to bring about the observed dislocations, Branco modified his original hypothesis.

Hypothesis 7 (Ries mountain) *The sudden elevation of the Ries mountain was accompanied by a (water vapor) explosion.*

This modified hypothesis, Branco argued, provides a better explanation of the chaotic debris. Moreover, he pointed out analogies to other sudden volcanic explosions without lava ejection.

Still, even Branco's modified hypothesis didn't satisfy Kranz (1911,14–52), who questioned the ability of a conventional volcanic eruption to release sufficient amounts of energy. As a result, Kranz proposed the so-called blasting theory.

Hypothesis 8 (Blasting) *Ground water entered a magma chamber situated in shallow depth, triggering a massive explosion.*

Kranz' theory gave a unified account of the diverse evidence. It explained, for example, the different items of polished and striated surfaces as resulting from the impacts of rock fragments which were catapulted by the explosion. According to Kranz' blasting theory, though, the Ries was conceived as a geologically unique phenomenon. Nonetheless, blasting theory became, gradually, the consensus view of geologists, and remained so until 1960.

It was the strangeness of the Ries which led Werner (1904) to surmise a radically different theory about the Ries' origination.

Hypothesis 9 (Impact) *The Ries represents the remnants of an impact crater.*

Werner likened the Ries to lunar craters, but failed to give arguments in favor of his hypothesis. Two further authors, Kaljuwee (1933) and Stutzer (1936), had advanced the impact theory before 1960. While Kaljuwee argued that the Earth had been subject to massive meteoritic bombardment in the past, whose traces cannot have been entirely erased from Earth's surface, Stutzer compared the Ries with the Barringer Crater in Arizona, noting significant morphological similarities. The impact hypothesis was nonetheless dismissed by the scientific community. This changed, however, radically and sustainably in 1960 once Shoemaker and Chao detected the rare mineral coesite, which was first discovered in the 1950s and which crystallizes only at very high pressures, in samples of rocks from the Ries. Both had previously found coesite in the Barringer Crater, as well. Since the extreme pressures required to form coesite cannot be reached by volcanic activities, this discovery was unanimously regarded as a successful verification of the impact theory, effectively closing the controversy.

The sketch of the Ries debate illustrates the kind of controversial argumentation this book's inquiry is going to analyze. Let us try to describe the lively debate in somewhat more abstract, argumentation-theoretic terms. The debate comprises different proponents who hold specific positions. These are modified in the light of new arguments introduced into the debate. Some of these arguments are intended to support a given position, others are set forth so as to attack opponents. Novel evidence enables the proponents to put forward ever new arguments. Proponents agree with each other to different degrees. Cotta and Voith, for example, holding Hypothesis 1 advanced by Caspers, concur, obviously, by and large with Schnitzlein and Frikhinger, who maintain only a slightly modified position (Hypothesis 2). In contrast, these proponents disagree more or less fundamentally with advocates of Schafhäutl's viscous underground theory (Hypothesis 3). As the debate evolves, and as proponents alter their positions, the overall agreement changes as well, resulting, as far as we can tell from our brief sketch, in phases with almost unanimous consensus (e.g., 1800–1840, 1920–1940), or outspoken plurality and dissent (e.g., 1900–1920). Besides mutual agreement, the overall truthlikeness, or "verisimilitude," of the proponent positions appears to vary too. So, some hypotheses seem to be closer to the truth (of course, relative to our current knowledge) than others. Gümbel's Hypothesis 5, for example, improves *objectively* upon previous volcanic hypotheses by attributing all suevite occurrences to one and the same source. Likewise, the refined Ries mountain theory (Hypothesis 7) gets much closer to the truth than hypotheses which posit glaciation but is itself outperformed by blasting theory (Hypothesis 8). In sum, there is nothing fundamentally obscure about assessing, in retrospective, the effects of controversial argumentation on the mutual agreement and verisimilitude of proponent positions. (That these concepts call, however, for more precise explications goes without saying.)

The general purpose of this book is to assess the instrumental consensual and veritistic value of controversial argumentation, in other words, to assess its consensus- and truth-conduciveness. One method for doing so would consist in providing, first of all, a *detailed* reconstruction and analysis of our example, specifying, in particular, how arguments are introduced into the debate, how proponents modify their beliefs in response, and how this affects the proponents' overall agreement, as well as the correctness of their convictions. Next, an equally detailed analysis would have to be carried out for dozens if not hundreds of further controversies so as to obtain a sufficiently broad sample of dynamic debate reconstructions. A statistical analysis of this sample could then teach us whether controversial argumentation is, in general, consensus- and truth-conducive and which specific argumentation strategies are particularly effective with regard to generating agreement and discovering the truth.

Obviously, that is a giant task, and is not going to happen, at least not here. It takes already one book to document the reconstruction and analysis of a single debate. The above paragraph hence outlines rather an entire research program than an investigation to be unfolded in a monograph.

Lacking a sufficiently large sample of dynamic debate reconstructions, which would allow us to learn from previous experience with different argumentative practices, we are going to generate our own tailored samples by simulating controversial argumentation. So, instead of studying real debates and their reconstructions, we investigate simulated debates and their automatically generated formal representations. This allows us, in principle, to scrutinize the effects of different argumentative practices in arbitrary many debates and therefore to identify their consensus- and truth-conduciveness accurately. Clearly, the simulation of controversial argumentation has to rely on an adequate model which incorporates the relevant aspects of debate dynamics. This model will be described, informally, in the following section. It extends the approach developed in Betz (2010) by a dynamic component and is carefully set forth in Chap. 2.

1.3 Modeling Controversial Debate

A fixed state of some debate is essentially characterized by (a) the arguments which have been uncovered and introduced so far and (b) the positions maintained by the debate's proponents. We assume that the arguments are—or are reconstructed and thence represented as—deductively valid inferences from some premisses to a conclusion. Arguments may support or attack each other, giving rise to a complex argumentation which we will refer to as a dialectical structure and which may be visualized as an argument map.

Given a dialectical structure containing arguments which mutually support and attack each other, we can identify a position, actually or potentially held by some proponent, with a truth-value assignment to the sentences which figure in the debate. We refer to a truth-value assignment to all the sentences which occur in the debate as a "complete position"; a partial position, in contrast, maps truth values to some of the sentences only. While assuming, throughout the following study, that proponents hold complete positions, we do distinguish, in some cases, so-called core and auxiliary beliefs, capturing the Lakatosian idea that proponents don't regard all sentences which figure in a debate as equally important.

The arguments advanced in a debate entail certain constraints a position ought to satisfy so that a proponent may reasonably adopt it. In addition to assigning equivalent sentences identical truth values, and contradictory sentences complementary ones, a position must, on account of deductive validity, consider a conclusion of some argument true, if its premisses are deemed correct. We shall call a complete position "dialectically coherent" if and only if it satisfies these constraints.

Since positions are identified with truth-value assignments, their mutual agreement can simply be assessed by counting the number of sentences to which two positions assign the same truth value. This gives us a simple metric on the set of all positions, and allows us to picture the set of coherent positions as a space in which the proponents (provided they hold dialectically coherent positions) are located.

When investigating the veritistic value of controversial argumentation, we stipulate that some position (truth-value assignment) is correct and represents the true truth-value assignment, that is, the truth. The truth is dialectically coherent (for no deductive valid argument has true premisses and a false conclusion) and is itself located in the space of coherent positions. Assessing a position's agreement with the truth yields a convenient way for gauging its truthlikeness (verisimilitude).

The background knowledge shared by the debate's proponents (*endoxa*) represents an additional characteristic of a given state of debate. We model background knowledge as a partial position which is accepted by all proponents; more precisely, the complete proponent positions necessarily agree with the background knowledge. Moreover, we assume that the background knowledge is (1) constant and (2) correct, that is, agrees with the truth. In other words, we don't consider the case where proponents systematically err with respect to background assumptions.

So far, we have merely given a static account of a debate, which focuses on some fixed state of a controversial argumentation, taking a single snapshot.

Clearly, this framework could be applied to reconstruct consecutive states of a real debate, which would provide a dynamic picture of how the individual states evolved into one another. Yet, a simulation of debate dynamics (in contrast to its mere reconstruction) requires, in addition, that we model the way a given state of a debate triggers a further one. In particular, we have to describe how the two most important constituents of some state of debate, that is, the dialectical structure (comprising the arguments advanced so far) and the proponents' positions, evolve. Accordingly, we must detail an *argument construction mechanism* and a so-called *update mechanism*. The individual simulation studies documented in this book's chapters vary, primarily, in regard to the specific argumentation and update mechanism they presume. We shall roughly summarize and categorize these various assumptions in the following.

The most simple *argument construction mechanism* posits that new arguments be devised randomly, that is, that the premisses and conclusion of a new argument be drawn randomly from the set of all sentences which pertain to the debate. Consequently, arguments are not purposefully contrived by proponents and relate only coincidentally to the positions proponents maintain. According to random argument construction, arguments—previously unseen inferential relations—are rather discovered than designed.

More sophisticated argument construction mechanisms assume that arguments are introduced by a specific proponent who follows a certain argumentation rule, taking the positions, held by the debate's participants, into consideration. An important aspect which distinguishes such argumentation rules is the relative importance attached to the position of the proponent who advances the new argument versus her opponents' positions. We may thus distinguish argumentation rules which prescribe to introduce an argument that (a) backs the proponent's position (the conclusion is maintained as true by the proponent) or (b) criticizes an opponent's position (the conclusion is denied by an opponent). Likewise, we can discriminate between rules which demand that the premisses of a new argument be accepted (a) by the proponent who advances the argument or (b) by one of her opponents. These distinctions give rise to four basic types of argumentation strategies:

Fortify: An argument satisfies the *fortify* rule iff the proponent who puts forward the argument considers its premisses and its conclusion true.

Attack: An argument satisfies the *attack* rule iff the proponent who puts forward the argument maintains its premisses while one of her opponents denies its conclusion.

Convert: An argument satisfies the *convert* rule iff the proponent who puts forward the argument maintains its conclusion while one of her opponents accepts its premisses.

Undercut: An argument satisfies the *undercut* rule iff an opponent of the proponent who puts forward the argument denies its conclusion while conceding its premisses.

Whereas arguments that satisfy *fortify* and *attack* take off from the proponent's convictions, *convert* and *undercut* urge proponents to base new arguments on premisses accepted by their opponents. The latter strategies are opponent-sensitive, whereas the former ones may be characterized as self-centered. Moreover, by prescribing to criticize an opponent's position, *attack* and *undercut* are more aggressive than, respectively, *fortify* and *convert*.

Besides random argumentation, the four argumentation rules, or variants thereof, constitute the primary argument construction mechanisms studied in this book. This enables us to examine how opponent-sensitive and self-centered as well as more or less aggressive argumentation strategies affect the consensual and veritistic dynamics of debates.

A final type of argumentation strategy we consider can be employed by proponents who hold a core position plus further auxiliary beliefs. Such a core position possesses a specific degree of justification, or robustness, relative to a given state of debate. A core position's degree of justification can be precisely defined in the framework of the theory of dialectical structures (cf. Sect. 2.2), and proponents may hence introduce new arguments so as to maximize the degree of justification of their own core position. We will refer to this rule, which represents a self-centered strategy, as *maximize robustness*.

The argument construction mechanism specifies how a dialectical structure grows from one step in a debate to another. Likewise, the *update mechanism* describes how proponent positions evolve and, in particular, respond to arguments that have been newly discovered and introduced into the debate. In a nutshell, we assume that proponents hold and retain dialectically coherent positions, and try to minimize the number of belief revisions which are necessary to do so. To understand the dynamics of proponent positions, it is important to note that dialectical coherency hinges sensitively on the dialectical structure against which positions are assessed, and hence on the arguments discovered so far. More precisely, a position which is dialectically coherent given a state of debate might become dialectically incoherent as new arguments are introduced and proponents have to take account of further inferential relations. If a newly introduced argument renders the position maintained by some proponent dialectically incoherent, the proponent modifies her truth-value assignments so as to reestablish dialectical coherency with respect to the enlarged dialectical structure. We assume, in addition, that proponents are conservative in the sense of revising their convictions only reluctantly. Specifically, proponents minimize the individual revisions of truth-value assignments so as to regain a dialectically coherent position. Or, in other words, proponents update their position to the closest coherent one.

The closest coherent update mechanism underlies most of the simulations presented in this inquiry. Occasionally, however, we presume a slightly more sophisticated revision policy. As previously remarked, we distinguish, in some simulations, core and auxiliary convictions. Proponents are assumed to stick particularly vehemently to their core beliefs while being much more willing to modify their auxiliary convictions. This suggests the following modification of the simple closest coherent update mechanism: If the complete position held by a proponent is rendered dialectically incoherent, the proponent determines, in a first step, all coherent positions that agree maximally with her core convictions. In a second step, she chooses from those positions the one that displays the greatest overall agreement with her previous position. This lexicographic update mechanism will be employed whenever we distinguish core and auxiliary sentences.

This sketch of how we model debate dynamics clearly exposes some simplifications and therefore suggests obvious extensions. To begin with, there is no reason to assume that proponents maintain but complete positions. To withhold judgement in regard to some sentence may very well be a reasonable doxastic state. Moreover, that is what happens in real debates all the time. Accordingly, a first interesting extension of this investigation could posit that proponents hold but partial positions. This would trigger a corresponding modification of the debate dynamics, in particular, of the update mechanism, which must allow for retracting truth-value assignments to some sentences altogether as well as for extending one's partial position. Secondly, future research might loosen the assumption that (explicit) background knowledge is constant and correct. The externally fixed background knowledge might itself grow in the course of a debate; it is, moreover, not immune to revisions and might therefore vary considerably. A particularly interesting extension consists in studying the effects of background knowledge correction, that is, the revision of false yet previously universally shared beliefs. These brief remarks demonstrate that the investigation carried out in this book is, by no means, to be read as a final word. Rather, it paves the way for possibly even more interesting inquiries into the dynamics of controversial argumentation within the framework of the theory of dialectical structures.

1.4 Results Pertaining to Consensus-Conduciveness

In the following, we report and summarize the main results regarding the consensual value of controversial argumentation, which are derived, and discussed, in Part I of this book:

C1 (General Results) Controversial argumentation is, all things considered, consensus-conducive. Although the concrete agreement evolution in an individual debate seems to depend, mainly, on random factors, we may nonetheless discern substantial statistical differences between different argumentative practices.

C1.1 (Long Run) A controversial argumentation compels proponent positions to converge, eventually. This is, however, hardly surprising inasmuch as, in the long run, only one single position remains dialectically coherent. Different argumentative practices vary substantially with respect to the pace of this convergence.

C1.2 (Alienation) Controversial argumentation may very well, in particular during the initial phase of a debate, lead to the alienation of proponent positions, and undo coincidental agreement. Instead of generating agreement, controversy sparks dissent. This effect, too, depends strongly on the argumentative strategies employed by the proponents. It is, in line with (C1.1), inevitably reversed in the long run.

C1.3 (Global Agreement Versus Partial Consensus) There exists a trade-off between (a) increasing the overall mean agreement between *all* proponents in a debate and (b) prompting at least some proponents to agree fully. Debate evolutions which foster partial consensus (i.e., full agreement between some proponents) tend to slow down the global rapprochement of proponent positions.

C1.4 (The Space of Coherent Positions) The characteristic consensus dynamics of argumentative practices, such as, for example, the result (C1.3), can be explained in terms of how the corresponding argumentation shapes the space of coherent positions. In particular, the degree of fragmentation of the space of coherent positions—whether the remaining coherent positions, that is, are all closely related to each other, or form, on the contrary, distant and isolated opinion clusters—turns out to be of pivotal importance for the belief dynamics. The concept of the space of coherent positions is, in fact, the primary theoretical tool for understanding the consensus-conduciveness of argumentative practices.

C2 (Background Knowledge) The introduction of background knowledge into a debate fosters, very much as one would expect, the mutual agreement between proponents.

C2.1 (Multiplier Effect) The introduction of constant background knowledge accelerates the rapprochement of proponent positions. This is because, as the debate unfolds, ever more sentences are derived from the constant body of background beliefs. These sentences become, consequently, part of the effective background knowledge themselves and may, in turn, serve as a basis for the derivation of further statements. This multiplier effect drives the discernible speed up of mutual rapprochement.

C2.2 (Favorable Fragmentation) With a sufficiently broad body of background knowledge, the fragmentation of the space of coherent positions, which tends to obstruct mean agreement increase without background knowledge (C1.4), favors both the generation of partial consensus and the global increase of mean agreement, thus resolving the trade-off reported above (C1.3).

C3 (Argumentation Strategies) The consensus-conduciveness of specific argumentative practices varies widely. The most noteworthy differences pertain to self-centered argumentation rules on the one side and opponent-sensitive ones on the other side.

C3.1 (Self-Centered Argumentation) Self-centered argumentation strategies, that is, argumentation rules (such as *fortify* and *attack*) which stipulate that a proponent advances but arguments with premisses she accepts as true, are totally ineffective in generating agreement. Strategies which are in addition aggressive, recommending direct attacks against opponent positions (e.g., the *attack* rule), consistently destroy agreement in all phases of a debate and drive proponent positions systematically apart.[4]

---

[4]Note that this stands in no contradiction to result (C1.1) because at some point in a debate, there are typically no more arguments that satisfy the *attack* rule, and proponents have to resort to other strategies if the debate shall continue.

C3.2 (Opponent-Sensitive Argumentation) Opponent-sensitive argumentative practices, however, are highly consensus-conducive. So, using, as premisses of the arguments one introduces to back up one's position, statements which an opponent considers true represents the most effective way for generating agreement. This result underlines the importance of explicitly addressing opponents by taking their positions as starting points for new arguments.

C3.3 (Aggressiveness and Disagreement) Aggressive opponent-sensitive strategies, that is, extremely critical strategies such as the *undercut* rule, are, in general, less consensus-conducive than their non-aggressive counterparts (*convert*). Too much criticism and too many direct attacks seem to inhibit rapprochement. The less aggressive *convert* rule, moreover, allows for an apparently highly beneficial strategy: Before directly refuting an opponent position, potential back doors (adjacent fall-back positions) which are available to the opponent and which are farther removed from the proponent than the opponent's current position are closed (rendered incoherent). When the opponent position is, afterwards, directly refuted, the opponent is compelled to relocate towards the proponent. Our simulations identify this complex mechanism and demonstrate its consensual value.

C3.4 (Friends and Fundamentalists) The effectiveness of an argumentation strategy in generating consensus depends on whether the initial agreement with one's opponent is very high ("friend") or very low ("fundamentalist"). Thus, a sharply critical, aggressive opponent-sensitive rule is advisable when arguing with a fundamentalist. Frequent falsifications due to "internal critique" represent in fact the most appropriate means for overcoming extreme dissent (cf. Schleichert 1998, pp. 93–111). Massive criticism impedes, however, finding consensus when arguing with a friend. Minor disagreement, instead of being effectively resolved, is typically deepened by aggressive argumentation. Here, the less critical *convert* strategy is much more consensus-conducive than the *undercut* strategy.

C4 (Consensus Bias) Different argumentative practices do not only vary with respect to their consensus-conduciveness. The argumentation strategies employed by the proponents affect, in addition, the distance between the proponents' initial positions and the debate's final consensus.

C4.1 (Resilient Argumentation) The final consensus reached in a debate tends to be closer to the initial positions held by proponents who employ an opponent-sensitive argumentation strategy (i.e., the *convert* or *undercut* rule) than to the initial positions maintained by proponents who argue in a self-centered way (implementing the *fortify* or *attack* rule). Thus, a proponent who follows an opponent-sensitive argumentation rule does not only foster consensus but also benefits from the ensuing fact that the final consensus is, on average, biased towards her initial position.

C4.2 (Robust Core Positions) Proponent core positions with a high degree of justification at an early phase of the debate tend to be closer to the final consensus than core positions which exhibit a low degree of justification. Degree of justification (at an early state of the debate) correlates with a position's agreement with the final consensus. This is because the higher the degree of justification, the more robust the corresponding core position and the more flexibly can a proponent adapt her complete position to critical arguments without modifying her core beliefs. This result provides a first justification for why adopting a position with a high degree of justification is rationally desirable at all (see also T4.3 below).

C4.3 (Sensitivity of Indicator) Proponents who introduce arguments so as to maximize the robustness of their core position don't reach a consensus any faster than proponents who apply opponent-sensitive strategies. Yet, in debates where proponents maximize their robustness through argumentation, the accuracy of the degree of justification as an indicator of a position's agreement with the final consensus increases dramatically. In contrast, the correlation between robustness and agreement with the final consensus almost vanishes entirely if proponents pursue very aggressive and critical strategies. Robustness seems to be a highly sensitive indicator.

1.5 Results Pertaining to Truth-Conduciveness

This section summarizes our findings about the veritistic value of controversial argumentation. The following results are spelled out in much more detail in Part II of this book:

T1 (General Results) In toto, controversial argumentation enables proponents to track down the truth. Individual veritistic dynamics vary substantially from debate to debate and are mainly determined by random factors. Still, different argumentative practices give rise to specific mean verisimilitude evolutions, and can thence be characterized statistically.

T1.1 (Long Run) Proponent positions converge, in the long run, against the truth. As explained above (C1.1), this is not surprising. Argumentative practices differ, however, significantly with respect to the speed and timing of the verisimilitude increase.

T1.2 (Epistemic Deterioration) Controversial argumentation may trigger a temporary loss of, instead of a gain in, verisimilitude. Still, verisimilitude evaporates to a much lesser degree than mutual agreement in the course of a debate. That is because it is comparatively difficult to render proponent positions which are close to the truth dialectically incoherent.

T1.3 (Engine of Progress) Criticism, as Mill argued no less eloquently than prominently, is indeed the main driver of epistemic progress.[5] The pace at which

[5]In *On Liberty*, Mill defends freedom of speech on grounds of the epistemic virtues of controversial discussion."Complete liberty of contradicting and disproving our opinion, is the very condition which justifies us in assuming its truth for purposes of action; and no other terms can a being with human faculties have any rational assurance of being right." (Mill 2009, p. 60) "The steady habit of correcting and completing his own opinion," Mill details, "is the only stable foundation for a just reliance on it: for, being cognisant of all that can, at least obviously, be said against him, and

proponents approach the truth is largely determined by the frequency at which their positions are rendered incoherent (successfully criticized). Rendering a proponent position incoherent requires, however, that one pinpoints an internal inconsistency pertaining to a subset of the proponent's beliefs, not all of which must, as deductive logic has it, be true. The fact that not all sentences figuring in an alleged inconsistency may be true, whereas, of course, they may all very well be false, amounts to a small but nonetheless influential asymmetry, which assures that, on average, internal critique tends to target more false than correct beliefs, and thus prompts a proponent to modify her position for the better.

T1.4 (Consensual and Veritistic Value) The relationship between consensus- and truth-conduciveness is intricate. A highly truth-conducive practice is necessarily consensus-conducive, at least to a certain degree, for it impels proponents to assent, gradually, to one and the same position, the truth. Yet, consensus-conduciveness alone does not guarantee truth conduciveness, and can, in fact, prevent proponents from approaching the truth. Argumentative practices which are highly effective in promoting agreement tend to generate spurious consensus.

T1.5 (Space of Coherent Positions) As in the case of consensus-conduciveness, the degree of fragmentation of the space of coherent positions exerts a markable influence on a debate's veritistic dynamics and represents thus a pivotal explanatory variable. As a rule (with several notable exceptions, though), debates with a highly fragmented space of coherent positions display lower verisimilitude increase.

T2 (Background Knowledge) Background knowledge affects an argumentation's truth-conduciveness in similar ways as its consensus-conduciveness.

T2.1 (Multiplier Effect) Constant background knowledge does not simply increase the mean verisimilitude of proponents by a fixed amount but accelerates their approaching the truth, since ever more sentences can be derived from the constant body of background beliefs during a debate.

T2.2 (Favorable Fragmentation) With sufficiently many correct background beliefs, the fragmentation of the space of coherent positions turns out to be favorable rather than detrimental to an argumentation's truth-conduciveness. This effect, however, is less pronounced than with consensus-conduciveness (C2.2).

T3 (Argumentation Strategies) The veritistic value of an argumentative practice does not correspond, one-to-one, with its consensual value. A proponent's ability to track down the truth is determined by her own argumentation strategy as much as by her opponents'. We find that argumentative practices differ significantly in terms of their characteristic truth-conduciveness.

---

having taken up his position against all gainsayers—knowing that he has thought for objections and difficulties, instead of avoiding them, and has shut out no light which can be thrown upon the subject from any quarter—he has a right to think his judgement better than that of any other person, or any multitude, who have not gone through a similar process." (Mill 2009, p. 64) Criticism, though, has no intrinsic but merely instrumental epistemic value. "Such negative criticism would indeed be poor enough as an ultimate result; but as a means of attaining any positive knowledge or conviction worthy the name, it cannot be valued too highly [...]" (Mill 2009, p. 128).

T3.1 (Veritistic Value of Critique) As the advancement towards the truth is primarily driven by criticism (T1.3), proponents whose positions are frequently rendered incoherent exhibit a comparatively rapid verisimilitude increase. In consequence, it is the argumentation strategy employed by one's opponent, and this opponent's ability to advance critical arguments, which controls the pace at which one acquires more and more true beliefs. Proponents whose opponents argue in an aggressive and opponent-sensitive way (*undercut* rule) display, accordingly, the strongest verisimilitude rise. Opponents, in contrast, who don't address a proponent's position at all, arguing in a self-centered way, don't allow the proponent to improve her position. These findings, too, corroborate Mill's methodology of controversial debate, in particular his emphasis on being criticized by able opponents: "So essential is this discipline to a real understanding of moral and human subjects, that if opponents of all important truths do not exist, it is indispensable to imagine them, and supply them with the strongest arguments which the most skilful devil's advocate can conjure up" (Mill 2009, p. 108).

T3.2 (Veritistic Value of Plurality) Outstanding consensus-conduciveness and the inability to question (and give up) a reached consensus contributes to an argumentative practice's consensual value but tends to curtail its veritistic one. This is strikingly revealed by our simulations, where proponents who implement the *convert* rule fare poorly in terms of verisimilitude. Now, high initial disagreement and the employment of agreement-reducing strategies, side by side with consensus-conducive ones, can help to avoid the emergence and persistence of a spurious consensus and enable proponents to continue questioning their beliefs. Plurality, we find, is an instrumental epistemic virtue, and argumentative practices which explicitly cultivate it (in an, otherwise, extremely consensus-conducive climate) foster a debate's overall truth-conduciveness. Once more, Mill had it right: "[The] only way in which a human being can make some approach to knowing the whole of a subject, is by hearing what can be said about it by persons of every variety of opinion, and studying all modes in which it can be looked at by every character of mind" (Mill 2009, pp. 62–64).

T3.3 (Consensus First) Aggressive and opponent-sensitive argumentation (i.e., the *undercut* strategy) represents the most truth-conducive practice in dualistic (i.e., two-proponent) debates. This is, however, not the case if multiple proponents engage in a controversy. Instead of fervently criticizing the various proponent positions simultaneously, it is more efficient to generate a consensus, possibly a spurious one, in a first step, and to criticize the consensus position (by way of self-critique) in a second step. This more conciliatory strategy, it turns out, is, in sum, more truth conducive than an immediate criticism of the diverse proponent positions. A specific version of the *convert* rule has, consequently, a role to play in truth-seeking controversies, as well.

T4 (Veritistic Indicators) Because—on average, and irrespective of the argumentative practice employed—proponent positions approach the truth only in a relatively advanced phase of a debate and since, in addition, real debates (for lack of new arguments) often don't attain these advanced phases, it becomes a decisive question whether there are reliable methods for gauging the verisimilitude of proponent positions in an early stage of a debate. We may identify, accordingly, three veritistic indicators, whose characteristic properties are summarized below: consensus, stability, and degree of justification. Remarkably, these indicators suggest a novel, "dialectic" foundation of the two major methodologies which have been developed in philosophy of science, that is, falsificationism and verificationism.

T4.1 (Consensus) Consensus, for being possibly spurious, may obviously be a misleading indicator of truth. Still, a consensus which is reached not simply by two, but by at least five or six (independently arguing) proponents is typically a very good indicator of truth. In general, the greater the size of a consensus (in terms of proponents), the higher its expected verisimilitude. If the proponents who reach the consensus display substantial initial disagreement, the reliability tends to improve even further. The accuracy of consensus as an indicator of truth depends, moreover, on the specific argumentation strategy employed by the proponents. The more consensus-conducive the argumentative practice, the less reliable the indicator. In a highly critical controversy (proponents follow the *undercut* rule), however, even a two-proponent consensus represents a highly accurate indicator of truth, especially at an early stage of the debate. Thus, given the appropriate argumentative practice, consensus allows one to infer verisimilitude in a reliable way. Our inquiry hence confirms a very old, in fact ancient, methodological idea, which runs, for instance, already through Plato's dialogues.[6]

T4.2 (Stability) The stability of a proponent position can be measured in different ways—as agreement of the position with the proponent's initial position, or as relative frequency at which the proponent had to modify her previously held positions. No matter how one gauges stability, however, it yields a telling indicator of a position's verisimilitude at an early stage of a debate. The accuracy of stability as an indicator of truth depends on the argumentation strategies pursued by the debate's proponents. Specifically, the more critical the argumentation, the more accurate the indicator. With proponents who implement the *undercut* strategy, stability becomes in fact an extremely reliable indicator of truth. This allows us to make sense and to justify core tenets of a refined falsificationist methodology. In a modified account of his earlier views, Popper (1963), reaffirming the importance of submitting theories to severe tests (criticism), introduces the idea that the iterative process of continuous testing gradually increases the verisimilitude of our theories, namely of those which pass the successive batteries of tests. Here is

[6]Consider, for example, how Socrates, having attested to Callicles' goodwill, frankness, and (pivotally) critical competence, addresses the latter: "Well then, the inference in the present case clearly is, that if you agree with me in an argument about any point, that point will have been sufficiently tested by us, and will not require to be submitted to any further test. For you could not have agreed with me, either from lack of knowledge or from superfluity of modesty, nor yet from a desire to deceive me, for you are my friend, as you tell me yourself. And therefore when you and I are agreed, the result will be the attainment of perfect truth" (*Georgias*, 487).

a dialectic reformulation suggested by our inquiry's results: Theories (positions held by proponents) which remain stable in the face of critique are, on average, closer to the truth. Their ability to pass controversies unspoiled testifies to their verisimilitude. And criticism is crucial, as Popper rightly sees, because stability constitutes a revealing indicator of truth only on the condition that the debate is highly controversial and proponents argue in an aggressive and opponent-sensitive way.

T4.3 (Degree of Justification) The verisimilitude of a proponent's core position is, at an early stage of a debate, correlated with its degree of justification. Degrees of justification thus signal proximity to the truth. Holding a partial position with a high degree of justification is veritistically valuable. The correlation between degree of justification and verisimilitude is particularly strong if arguments are discovered randomly or introduced by proponents with a view to maximizing their positions' robustness. This relatively simple result seems to provide a justification of inductive modes of reasoning—understood as meta-reasoning on dialectical structures with probabilities interpreted as degrees of justification. As I've tried to show elsewhere, such a dialectic framework, in turn, licenses inference to the best explanation and confirmatory inferences in line with the hypothetico-deductive account of confirmation, besides Bayesian inferences and reasoning with precise probabilities (see Betz 2011c,d). With a view to apparent theoretical parallels to Wittgenstein's[7] and Carnap's[8] definition of logical probability, one may conceive these results as a *dialectic foundation of probability*.[9]

T4.4 (Methodological Trade-Off) The last two results seem to suggest that falsificationism and verificationism don't represent mutually exclusive methodologies but stand for alternative, yet equally viable ways to estimate the verisimilitude of hypotheses (proponent core positions) at an early stage of a controversy. Nothing seems to prevent one from using both stability and degree of justification as veritistic indicators. As both indicators are fairly accurate in random debates, this is principally possible. Yet, if one attempts to sharpen the accuracy of stability by stipulating that proponents argue in a highly critical way, the reliability of degree of justification as an indicator of truth is completely lost. If not a definite opposition, there seems to remain a certain trade-off between the two indicators, and, accordingly, between falsificationist and verificationist methodologies, because the accuracy of the indicators, and the reliability of the corresponding inferences, hinges sensitively on the argumentation strategies pursued by the proponents. In addition, this fact obviously complicates the application of these methods to real controversies.

---

[7] cf. *Tractatus*, 5.15.

[8] Carnap (1950).

[9] See also Betz (2010, 61). Let me clarify, however, that we have, of course, not solved Hume's problem. Far from giving a justification for induction in general, we rely, throughout this inquiry, on inductive methods, for example, in the form of basic statistical reasoning (applied to the debate ensembles).

1.6 Objections and Caveats

Before we put the approach pursued in this inquiry in perspective, relating it to other theories of rational belief dynamics in the subsequent section, we shall consider some general limitations of our approach and objections which may be raised against it. This will allow us to delineate the scope of our investigation and to announce important clarifications and caveats.

First, the proponents in our model are (unrealistically) rational. They revise their convictions but in the face of arguments, and their belief dynamics are merely determined by the inferential relations encoded in the dialectical structure. (If, however, these logical constraints underdetermine the belief revision, because there are several closest coherent positions, the proponents are indifferent and make a random choice.) Yet, we all know that our beliefs are shaped not only by arguments but by many other factors as well. Pride or arrogance might prevent proponents from taking novel arguments fully into account. Different kinds of attachment such as fondness or loyalty might cause us to lean towards the positions held by a dear proponent. Hostility or contempt, in contrast, might prevent us from conceding a point. The way a claim is framed and the way it is rhetorically presented might affect our inclination to agree or disagree. The frequency at which a statement is uttered, too, seems to influence our tendency to believe it. This illustrative enumeration raises the question whether it is admissible to ignore these factors when modeling controversies. I posit it is—as long as one reads our model as a normative one. Thus, we are not trying to give an empirically adequate account of real debates and opinion dynamics but trying to assess certain fundamental properties of debates of an ideal type, namely, their consensus- and truth-conduciveness. Rather than allowing for, say, accurate predictions of real belief formation, these studies enable us to conclude how we *should* try to argue if we are interested in achieving consensus or in finding the truth by rational argumentation.

Second, our model has it that the proponents, when pursuing an argumentation strategy, invariably succeed in designing and introducing tailored arguments. But is this even possible? Can we construct arguments at will? Or, more precisely, can one always find, given some sought-after conclusions and a set of potential premisses, a deductive argument which inferentially links some of the premisses with one of the wanted conclusions? At first glance, the answer is, plainly, no. Consider, for instance, the case where the potential premisses on the one hand and the desired conclusions on the other hand are logically independent. Then no deductive argument whatsoever links premisses and conclusion in an appropriate way. But that seems to imply that real proponents simply cannot mimic the ideal types of controversies we are studying, even if they tried hard, because the argumentation rules, so successfully employed by our model proponents, face real limitations, even logical ones. This amounts to a severe challenge, and there are two ways to address it. The first rebuttal of the challenge stresses that, while the model assumption might be unrealistic indeed, it is not that far off as the objection makes believe. Our actual capacity to design arguments is not negligible. So, while the no-failure assumption, that is, the supposition that proponents never fail in introducing an argument which satisfies certain intended specifications, clearly overstates our abilities, the opposite no-success assumption, which presumes that proponents never find arguments that satisfy an argumentation rule, is equally mistaken. If a person, engaged in a debate, attempts to support her own position, and tries to advance corresponding arguments, she usually succeeds in doing so, at least from time to time. Likewise, if one attempts to criticize an opponent, searching for arguments that attack the opponent's claims, this is far from being a hopeless endeavor and leads, albeit not always and invariably, to the introduction of suitable arguments. In sum, I concede that the no-failure assumption, built into our model, is unrealistic and, strictly speaking, unattainable. Yet, our argumentative practice teaches us that we can successfully design arguments to some degree. With a view to identifying and contrasting the effects of different argumentation strategies (which, admittedly, can only be imperfectly applied in real debates), the no-failure assumption seems to me, nevertheless, an appropriate first-order approximation, since it allows for the most distinct assessment of the (purely applied) argumentation rules. Let us now turn to the previously announced, second rebuttal of the objection against the no-failure assumption. It requires that we step back for a moment and reflect on different interpretations of our simulated dialectical structures. Recall that real debates, no matter whether they unfold in an oral or a written exchange, don't consist in deductively valid arguments and don't explicitly realize a dialectical structure. It takes a substantial amount of analysis and reconstruction to transform a raw argumentation into well-formed arguments and a uniform dialectical structure. The arguments in such a reconstructed dialectical structure are deductively valid by virtue of the reconstruction, and relative to a given logic, the reconstruction logic, which the analysis rests upon. The first and obvious interpretation of our simulated dialectical structures is to understand them as debate reconstructions, containing arguments that are valid inferences relative to some reconstruction logic, with all premisses made explicit. It is within this first interpretation where the doubts about the no-failure assumption (Can one always establish, given the reconstruction logic, sought-after inferential relations between some sentences?) arise. Yet, an interpreter, reconstructing a debate, has not merely some leeway in choosing the reconstruction logic but may also decide on the reconstruction's "degree of explicitness." Thus, she may judge that, for example, mathematical principles, which are not warranted by the reconstruction logic itself but which are nevertheless universally agreed upon, might be omitted and don't have to be explicitly stated as premisses. Likewise, in a reconstruction of a debate about a company's future investment strategy, physical knowledge might equally be taken for granted and might hence not be presented explicitly in the reconstructed arguments. In Sect. 1.3 above, we introduced a direct representation of background beliefs, namely as fixed truth-value assignments. Here, we spot a second, indirect representation of background knowledge: Background beliefs might simply be omitted in the reconstructed debates, so that only premisses which don't belong to the body of background beliefs are made explicit. Now, this yields a second interpretation of simulated dialectical structures: The modeled arguments stand for deductively valid inferences (relative to the reconstruction logic) which derive a conclusion from the explicitly stated premisses *and* the (implicit) global background beliefs. With this interpretation, designing arguments, which take off from given premisses and back a specified conclusion, becomes a completely different task: It doesn't merely consist in scrutinizing and crunching the inferential relations between the sought-after premisses and the wanted conclusion but comprises, primarily, the search for appropriate background beliefs (not made explicit in the argument) together with which the explicit premisses imply the conclusion. Clearly, this is much less difficult a task than finding a logical relation between a couple of statements held by a proponent, especially if the proponents have, implicitly, a lot of sufficiently diverse beliefs in common. In that case, a proponent can typically find an argument which inferentially links some premisses—on the basis of further implicit and shared background assumptions—with a wanted conclusion and which dovetails with a given argumentation rule. The idea that proponents can construct arguments which fit a given argumentation rule becomes even more plausible, if we consider that, in some controversies, proponents can *generate* shared background beliefs, for example, by collecting specific observational data or carrying out experiments. So, the second rebuttal stresses that the no-falsity assumption ceases to be unrealistic, and arguments can be successfully designed indeed, if the (implicit) body of background knowledge is sufficiently broad.

Third, the simulations, in particular those which are supposed to study the veritistic dynamics of debates, stipulate that some given position is correct. This position, the truth, is chosen randomly (at the very beginning). But, then, don't the simulations merely demonstrate, for example, how quickly proponents approach an arbitrarily selected position and not how rapidly they find the truth? In the end, the truth is not simply a randomly chosen position, is it? Admittedly, the specific setup of the simulations is prone to triggering confusions and worries of this kind. So let me try to clarify, with the help of an analogy, why choosing the true position randomly when initializing the simulation is not only unproblematic but even necessary in order to obtain meaningful results at all. Picture an engineer who has designed a machine which is supposed to test freshly fabricated footballs. Specifically, the machine scans a football's skin by moving a highly accurate and well-tested sensor over its surface. The sensor's path is determined by a complicated algorithm. Yet, before the procedure is employed at a large scale, the manufacturer urges to assess its reliability, that is, its ability to track down fissures in a football's hull. With the sensor itself being well tested, the crucial ingredient is the algorithm that prescribes the sensor's path. Thus, the engineer sets up a simulation which represents (1) a damaged football and (2) the sensor moving over the football's surface according to the corresponding algorithm. This simulation determines, for a given initial position of the sensor and the location of the fissure, whether the sensor, controlled by the algorithm, moves over the fracture, or not. Based on sufficiently many simulations, with varying initial conditions, the engineer is in a position to assess whether the algorithm reliably prompts the sensor to move over the fissure. Or, to put it—with a view to our analogy—differently, by way of simulating the procedure, the engineer assesses its instrumental value with respect to tracking down the rupture. Now, it is not merely perfectly fine, but even crucially required that the simulations assume that the fracture be located on a randomly chosen spot on the football's surface. For the engineer wants to assess the procedure's reliability in regard to finding any fracture, no matter where it is located. To assume, in contrast, that the football is damaged at a very specific spot, say 5 cm south of the sensor's initial location, is obviously not very helpful, since it doesn't evaluate the procedure's ability to detect damages at different places. In close analogy, it is crucial for our simulations that we don't make any (arbitrary!) assumptions about the particular location of the truth within the space of coherent positions. So, by presuming that the truth is an arbitrary (randomly chosen) position, we avoid, in fact, fatal arbitrariness and ensure that our simulations assess the veritistic value of controversial argumentation, that is, its instrumental value with respect to tracking down the truth.

Fourth, our inquiry is, critically, language-relative. Thus, we assess the consensus- and, in particular, the truth-conduciveness of controversial argumentation under the assumption that proponents speak a common and unvarying language. More seriously, we frame the truth by identifying it with a fixed position, within a given conceptual scheme. But then, one may object, we don't assess the ability of argumentative practices to track objective, in the sense of language-independent truth.[10] I suggest that the appropriate response to this alleged *reductio* consists in embracing its conclusion. Indeed, we suppose that, throughout a debate, proponents speak one and the same language. And we study which argumentative strategies allow them to achieve their epistemic aims, given the linguistic and conceptual means they have. Furthermore, the concept of truth we posit is, in fact, not a metaphysical one. This book's investigation doesn't attempt to demonstrate how to reach, by way of controversial argumentation, a mind- and language-independent, eternal and infallible truth. Hence, it doesn't address the fundamental skeptical challenge, which vexes traditional epistemology, either. In contrast, our investigation into the veritistic value of argumentation is based on an internal rather than metaphysical realism. It builds, accordingly, on a language-relative (yet nonetheless objective) notion of truth, as introduced by Carnap (1956), later defended by Putnam (1981), and, I take it, proficiently wrapped up by Kitcher (2001). As a consequence, we assess the ability of proponents, who engage in a debate, to track down the truth given the language they speak. Such an ability, however, does not imply that the verisimilitude of proponent positions thus attained is invariant to translations into other languages. So, let us assume, for the sake of illustration, that a controversial debate has led proponents reliably towards the truth. Assume, in addition, that the proponents' positions have to be translated, subsequently, into a new language because the proponents have decided to modify their conceptual scheme substantially. Now, the translated proponent positions, in spite of having emerged from an argumentation (though in a different language), might be completely wrong. In other words, the

[10] A similar objection is advanced, more generally, against explications of verisimilitude (cf. Oddie 2008).

presumed ability of controversial debate to track down the truth does not guarantee that proponent positions retain a high verisimilitude once the underlying conceptual scheme is changed. This is important to notice since it draws a relevant limitation to our inquiry's scope: Thus, we disregard, and exclude from our investigation, far-reaching conceptual change, which plays, for instance, an important role, as Kuhn (1962) famously argues, in some ("revolutionary") scientific controversies.

Fifth, the evaluation of controversial argumentation, and hence our results about its consensual and veritistic value depends not only on the common language spoken by the proponents but on the reconstruction of the natural language argumentation as well. Since a precise reconstruction of a debate is seriously underdetermined by the speech acts which the proponents actually advance and since, more specifically, an interpreter, when reconstructing an argumentation, may choose, more or less at liberty, how to individuate single premisses, thereby determining the number of premisses per argument as well as the number of sentences in the dialectical structure, the assessment of a debate within the framework of the theory of dialectical structures appears to be, by and large, arbitrary. All the crucial evaluative variables seem to hinge on the interpreter's subjective choice: the degree of agreement between proponents, the verisimilitude of a position, a debate's inferential density, a partial position's degree of justification, the stability of a proponent position, etc. And this seems to imply that not the argumentative practices but rather the way a debate is interpreted determines whether the proponents have reached consensus or attained the truth. This objection doesn't, in the first place, criticize our simulation studies but questions the applicability of our findings to real debates. Now, the underdetermination of a detailed debate reconstruction is clearly a hermeneutical challenge, yet I doubt it undermines our investigation. Let's assume, for the sake of the argument, that the dialectic evaluation of a debate depends in fact sensitively on arbitrary choices of the interpreter. This alone does, however, not interfere with a meaningful assessment of how evaluative variables (such as agreement, degree of justification, etc.) have evolved during one and the same debate or are correlated with each other—provided the arbitrary hermeneutic decisions are not varied in the reconstructions of the debate's consecutive states. If, for example, the interpreter reconstructs a given reason as an argument containing three premisses in the initial phase of the debate, she must reconstruct it in the same way in later phases. What is, admittedly, obstructed by hermeneutic underdetermination is a sound interdebate comparison of evaluation results, such as, for instance, juxtaposing the verisimilitude of partial positions in two different debates. Still, an assessment of the consensual and veritistic value of argumentative practices *relative to* the corresponding debate's starting point is all we are aiming at in this inquiry. So we examine, for example, which argumentation strategies tend to improve a proponent's veritistic situation in the course of a debate. And this sort of inquiry is not threatened by hermeneutic underdetermination. An analogy may clarify the rebuttal even further. Consider climatologists who try to assess the impact of a volcanic eruption in the nineteenth century on worldwide surface temperatures. The scientists base their investigation on temperature records from dozens of meteorological stations on different continents. Unfortunately, though, the stations used instruments which were not calibrated against each other, and the measurements of the different stations cannot be directly compared with each other, in consequence. This does, however, not interfere with a meaningful assessment of each station's temperature record—provided the stations didn't modify their instruments (within the relevant period). So for each individual station, the temperature effect of the volcanic eruption might very well be gauged. Moreover, by normalizing the data record with respect to the temperature before the eruption, the *relative* global mean effect of the eruption (e.g., −0.5 K) can be estimated. By analogy, we can assess the instrumental consensual and verististic value of different argumentative practices, even if evaluative variables were, in absolute terms, hardly comparable across different debates.

Sixth, by assuming that all arguments contained in a dialectical structure are deductively valid, we seem to pay no attention to inductive reasoning, which undeniably plays a crucial role in real debates and which certainly contributes to the consensual and veritistic value of controversial argumentation. This objection raises, of course, not only a challenge for our inquiry but for any approach in argumentation theory which subscribes to the principle of deductivism, that is, the view that all arguments, including the allegedly inductive ones, can and, ultimately, should be reconstructed as deductively valid inferences. I'm certainly not able, in the remainder of this section, to defend this view in depth. Such a defense, I suppose, requires to go through all types of so-called inductive arguments (e.g., reasoning by analogy, inference to the best explanation, statistical inference, enumerative induction), which allegedly resist a charitable (!) deductive interpretation, and to suggest, for each such type of argument, how to reconstruct it, deductively, in an appropriate way. In the following, however, I'd merely like to highlight (a) two different general reconstruction strategies which typically prove useful when reconstructing inductive reasoning, and to outline (b) a further possibility for embedding inductive modes of reasoning in the framework of the theory of dialectical structures. Note, ad (a), that every inductive argument relies on a specific inductive inference rule. An inductive inference rule has characteristically a different status than a logical one: It doesn't hold necessarily on account of certain logical constants, it may only be applicable as long as certain conditions are shown to prevail, and it might cease to be a sound inference rule as soon as new evidence emerges. Now, one straightforward strategy for reconstructing an inductive argument as deductively valid consists in making the inference rule, plus its additional applicability conditions and restrictions (e.g., ceteris paribus or total evidence clauses), explicit by stating it as a premiss of the argument. Consider, as an example, the following reasonably good inductive argument:

(P1) Tara is Indian.
(P2) Most Indians (>80%) are Hindi.
(C) Thus, Tara is Hindu.

A charitable deductive reconstruction makes the underlying inference rule explicit and adds the following additional premisses:

(P3) If (1) most F are G, (2) a is F, and (3) a being F represents our total evidence relevant to the question whether a is G or not, then a is G.

(P4) Tara being Indian represents our total evidence relevant to the question whether she is Hindu or not.

By adding (P3) and (P4), we obtain a deductively valid, monotonic argument. In particular, a defeat of the original inductive argument, such as learning that Tara lives actually in the region of Punjab, ca. 90% of whose inhabitants are Muslims, does not miraculously undermine the inference anymore but can now be explicitly related to premiss (P4), which becomes false as the new evidence surfaces. Furthermore, we may reconstruct the modified inductive inference, which makes use of the novel evidence, as follows:

(P1) Tara is an Indian living in Punjab.
(P2) Most Indians living in Punjab are Muslims.
(P3) If (1) most F are G, (2) a is F, and (3) a being F represents our total evidence relevant to the question whether a is G or not, then a is G.
(P4) Tara being an Indian living in Punjab represents our total evidence relevant to the question whether she is Muslim or not.
(C) Thus, Tara is Muslim.

Besides making the underlying inductive inference rule explicit, qualifying the conclusion of an argument represents a further valuable maneuver when reconstructing inductive arguments. More specifically, it might be necessary to insert probabilistic or epistemic operators so as to obtain plausible premisses and a charitable reconstruction. Likewise, our illustrative reconstruction might be further improved along the following lines:

...
(P3') If (1) most F are G, (2) a is F, and (3) a being F represents our total evidence relevant to the question whether a is G or not, then it is *likely/very likely/reasonable to accept/permissible to assume in further arguments...* that a is G.
...
(C') Thus, it is *likely/very likely/reasonable to accept/permissible to assume in further arguments...* that Tara is Muslim.

Demonstrating that inductive arguments can in fact be reconstructed as deductive inferences in a charitable way represents an effective rebuttal of the objection to deductivism. Yet, ad (b), a further and in some sense even more interesting rejoinder embeds inductive reasoning within the theory of dialectical structures not merely by interpreting these arguments as deductive inferences but by showing that inductive arguments can be understood as meta-inferences on dialectical structures. Thus, I have tried to explain, in separate articles, how (1) inductive reasoning in line with the hypothetico-deductive account of confirmation (Betz 2011c) and (2) inferences to the best explanation (Betz 2011d) may be understood as meta-syllogisms on a given dialectical structure. Moreover, by establishing that a partial position's degree of justification represents a significant indicator of verisimilitude (see T4.3 above), this inquiry strongly supports those results and contributes to a dialectic foundation of inductive inferences based on degrees of justification. All this dismantles the fear of our inquiry not fully and adequately accounting for inductive modes of reasoning.

1.7 Putting the Approach in Perspective

The endeavor to model the dynamics of belief change, and to understand the rationality thereof, is far from being novel. The investigation carried out in this book thence relates to a couple of alternative approaches in epistemology, philosophy of science, logic, and artificial intelligence, which attempt to explain the dynamics of rational belief formation and revision.

A major dimension along which these approaches can be ordered is the degree of logical competence which agents are assumed to possess according to the corresponding approach. One may, for example, assume that agents are logically omniscient, being aware not only of all inferential relations within a given set of sentences but even of all logical implications some sentence carries. This amounts to maximal logical competence and represents an extreme assumption in the spectrum we are considering. At the opposite side of this spectrum lies the presumption of minimal logical competence, or, as we shall call it, "logical ignorance." Agents are (modeled as) logically ignorant if they don't take account of any inferential relations between their convictions when revising their beliefs. This is in particular the case if a model of belief change doesn't represent inferential relations between sentences in the first place.

Models which presume that agents are logically omniscient comprise epistemic logic (Fagin et al. 1995), including dynamic extensions (van Ditmarsch et al. 2007) and theories of belief revision (Hansson 1999), in particular the so-called AGM model (Alchourron et al. 1985; Gärdenfors 1988), as well as, though to a lesser extent, argumentation frameworks as developed in artificial intelligence (Chesñevar et al. 2000; Prakken and Vreeswijk 2001; Bench-Capon and Dunne 2007).

Epistemic logic extends first-oder predicate logic by a knowledge operator K_i, allowing for the logico-semantical analysis of statements about an agent i's knowledge and for the expression of epistemic principles in the corresponding formal language. The syntactic calculus of epistemic logic is complemented by a possible world semantics (as proposed by Hintikka 1962) to the effect that $K_i p$ is interpreted as "p holds in every possible world which is compatible with what agent i knows." But as (1) a logical truth holds in every possible world, and (2) if p holds in some possible world then all its logical implications hold in that very world as well, we have $K_i p^*$ (with p^* being an arbitrary logical truth) and $K_i p \implies K_i q$ (with q being an arbitrary logical consequence of p) for every agent i. In other words, according to epistemic logic, agents are logically omniscient and hold deductively closed knowledge claims (see also Fagin et al. 1995, pp. 333–337; Hendricks 2006, p. 98).

The AGM model, named after its original authors Carlos Alchourrón, Peter Gärdenfors, and David Makinson, represents an agent's beliefs as a set of sentences in some formal language. Belief revision theories in the tradition of the AGM model study the principles of how an agent's overall belief set ought to change given (1) the acquisition of some new belief (*expansion*), the dismissal of some previously held belief (*contraction*), or the replacement of previously held beliefs by new ones (*revision*). Now, it represents a fundamental assumption of this approach, which seems to be required, as Hansson (2009) notes, in order to carry out an interesting formal treatment in the first place, that an agent's belief set be closed under logical implication. That is, agents are assumed to be logically omniscient. The AGM model has been used, recently, to investigate whether and under which conditions belief revision increases the verisimilitude of an agent's beliefs (see the special issue of *Erkenntnis*, in particular Kuipers and Schurz 2011). The ongoing research effectively brings together AGM theory and the program of (logically) explicating the concept of verisimilitude (cf. Niiniluoto 1998; Oddie 2008). Moreover, rather than simply trying to pin down precisely the notion of truthlikeness, such investigations take on the methodological challenge as formulated, for example, by Zamora Bonilla (1992, 2000), namely, to spell out how (i.e., through which methods) the verisimilitude of a belief set can be increased. Yet, results by Niiniluoto (2011) suggest that belief revision does not necessarily help agents to approach the truth. In general, these specific approaches, while being driven by a similar research interest than this study, remain committed to the assumption of logical omniscience.

Researchers in artificial intelligence (AI), taking Reiter's default logic as a starting point (Reiter 1980), have developed, in recent decades, a variety of approaches to modeling complex argumentation. In AI, controversies are typically analyzed as "argumentation frameworks." Although these theories of argumentation frameworks are not primarily concerned with the rational *dynamics* of belief change but attempt to model, rather, a static knowledge base in terms of its arguments, we shall nevertheless briefly consider them here on account of their resemblance with the theory of dialectical structures (see also Sect. 2.1). Some theories of argumentation frameworks, specifically those in the tradition of Dung (1995), are not explicitly based on formal logic at all, so it is not fully correct to say that these models assume agents to be logically omniscient in a strict sense. Still, I maintain that they (implicitly) assume agents to be logico-argumentatively omniscient, namely inasmuch as the corresponding evaluation procedures suppose that all potentially relevant arguments be taken into consideration when assessing a controversial claim. Let me illustrate this diagnosis with respect to the highly influential approach of Dung (1995). Dung takes it that "[for] a rational agent *G*, an argument *A* is acceptable if *G* can defend *A* (from within her world) against all attacks on *A*." (Dung 1995, p. 326) Yet, unless the argumentation framework contains every argument that can possibly be advanced at all (and unless an argument is, consequently, inacceptable if and only if it is simply not possible to defend it against an attack), Dung's explication of the fundamental notion of acceptability appears to be inappropriate, for a rational agent with limited cognitive capacities might, even in the face of undefeated counterarguments against her claim, stick to her position by simply saying that she does not accept the counterargument (denies one of its premisses), without being able, as of today, to back up that refutation with an extra argument. Note that a similar assumption is also built into the model developed by Besnard and Hunter (2008): Assuming that an argument which is not attacked has to be conceded by a rational proponent (p. 108) makes only sense insofar as the argumentation framework contains all relevant arguments which can possibly be advanced in the debate. In sum, the evaluation procedures established by AI models of complex argumentation seem to suppose that agents are, in a broader sense, logically omniscient as well.

Unlike the approaches considered so far, other models of rational belief dynamics don't represent, at least not explicitly, inferential dependencies between an agent's beliefs at all and hence seem to assume that agents are logically ignorant. The archetypal models of rational consensus formation and opinion dynamics developed by Lehrer and Wagner (1981), on the one hand, and by Hegselmann and Krause (2002), Hegselmann (2004), on the other hand, belong to this type (for a review which focuses on veritistic opinion dynamics see also Douven and Kelb 2011).

Both the Lehrer–Wagner and the Hegselmann–Krause models represent an agent's belief as a real number in the interval [0;1] and assume, at least in their most basic variants, that an agent's belief at step $t+1$ is fully determined by that agent's as well her peers' beliefs at step t. Now, the models diverge in terms of how the belief of agent i and the beliefs of i's peers are aggregated so as to yield the updated belief of i. In the Lehrer–Wagner model, each agent i assigns constant real numbers to her peers and herself, assessing the agents according to their alleged expert status. An agent i's new belief is then defined as the weighted average (based on the weights assigned by i) of all agents' previous beliefs. Lehrer and Wagner (1981) demonstrate that the agents' beliefs necessarily converge (if weights are greater than 0), and postulate that the resulting opinion dynamic represents a rational procedure for consensus generation. In the Hegselmann–Krause model, an agent i doesn't consider the opinions of all other agents when updating her belief but merely those peer beliefs which fall within a certain ε-interval ($\varepsilon > 0$) around i's belief. Agents, according to the intended interpretation, merely take those peers into consideration whose opinions are not too far off. Agent i's new belief is the plain average of all opinions that fall in her ε-interval. Hegselmann and Krause (2002) simulate the ensuing opinion dynamics and show, for example, that the size of the confidence interval (ε) crucially affects whether the agents settle on a consensus position or not. Extending the basic model, Hegselmann and Krause (2006) stipulate that some real number is the correct opinion and assume that a few agents possess the ability to track the truth: The beliefs of truth trackers are both affected by the peers within the corresponding confidence interval and attracted by the truth. Although this extension provides interesting new results, it doesn't amount to an *explicit* inclusion of inferential dependencies in the model. In sum, the Lehrer–Wagner as well as the Hegselmann–Krause model disregard inferential relations which hold between the agents' beliefs altogether, and conceive agents, accordingly, as logically ignorant.

Extending the Hegselmann–Krause model, Riegler and Douven (2009) represent an agent's belief state by a binary evaluation of a propositional basis (rather than a single real number). The modified model, unlike the original one, hence contains a pivotal element of the theory of dialectical structures. The specific propositional basis employed by Riegler and Douven (technically: the canonically ordered set of state descriptions Riegler and Douven 2009, p. 150) allows even to encode inferential relations between different sentences. Consequently, the extended model comprises a rich representation of opinion sets. As a drawback, however, Riegler and Douven (2009) have to assume that belief states of agents be closed under logical implication. In other words, their modification of the Hegselmann–Krause model relies on the assumption of logical omniscience.

The theories of belief dynamics which presume agents to be logically omniscient on the one side, and the models of opinion dynamics which don't represent inferential dependencies at all on the other side constitute opposite poles of a spectrum of approaches to modeling rational belief dynamics and are both characterized by equally extreme (and, on the face of it, unrealistic) assumptions about the logical competence of agents and its role in rational belief formation. That is precisely what sets the model unfolded in this book, which falls well in between these two extremes, apart from previous approaches. For we assume, on the one hand, that agents are not logically omniscient. Instead, they consider merely the inferential relations discovered so far (i.e., the arguments explicitly introduced into a debate), when inspecting and, if necessary, revising their beliefs. But this is of course, on the other hand, far from presuming that agents don't consider any logical dependencies whatsoever when updating their beliefs. Agents, as modeled by the theory of dialectical structures, are anything but logically ignorant. By acknowledging the actual cognitive limitations of agents who engage in an argumentation, our approach can be understood as a bounded rationality model (e.g., Simon 1982) of belief dynamics.

The previous remarks, however, are not supposed to discard the alternative approaches simply on the grounds that they rely on specific idealizations. Models that assume logical omniscience, for instance, study how ideal agents should form and modify their beliefs (likewise, Levi (1991, p. 8) conceives these models as analyzing an agent's commitments rather than her consciously held beliefs), possibly yielding significant epistemological insights which might, in turn, bear on our everyday epistemic practices. Moreover, these models constitute, obviously, adequate representations of computational multiagent systems and give rise to valuable applications in artificial intelligence and computer science. The Lehrer–Wagner and Hegselmann–Krause models, however, representing agents, ostensibly, as logically ignorant, can be understood as highly aggregated models of opinion formation. By concentrating on the macrodynamics of belief revision, they deliberately disregard detailed (micro) argumentative processes and seek to capture dialectic mechanisms through (1) the general opinion-averaging (assuming that arguments increase peer agreement) and (2) the truth-tracking procedure (assuming that arguments increase verisimilitude). From this perspective, the approach unfolded in this book can be interpreted as a model of the micro- and mesodynamics of rational debate, thus complementing the Lehrer–Wagner and Hegselmann–Krause models, which attempt to represent the corresponding macrodynamics.

Still, notwithstanding other fruitful applications of models with logical omniscience or ignorance, I take it that models of this type provide in any case poor representations of the *detailed* dynamics of rational debates and of the belief change which is triggered by controversial argumentation: Rational proponents who engage in a debate undeniably adjust their position in the face of new arguments, without, however, being logically omniscient (which would render the entire *process* of argumentation, that is, the introduction of new arguments, superfluous and the corresponding real-world *practice* incomprehensible).

In the remainder of this section, we discuss two further approaches to modeling doxastic dynamics, which resist a straightforward subsumption under the opposite types of alternative theories previously considered. These approaches are, first, Paul Thagard's theory of explanatory coherence and (scientific) controversy and, second, theories of judgement aggregation, which constitute a lively research area, bringing together economists, sociologists, political scientists, scholars of law, computer scientists, and philosophers alike.

In his book *Conceptual Revolutions*, Paul Thagard develops a theory of explanatory coherence with a view to understanding the dynamics of scientific controversies (Thagard 1992, Chap. 4). Thagard considers propositions which state (1) the available observational evidence and (2) the proposed hypotheses at a given state of debate. He represents several relations which may hold between these propositions. Pivotally, his model maps explanatory relations between tuples of hypotheses on the one side and observational statements on the other side. Thagard specifies seven general principles which allow one to translate the explanatory links between propositions into a symmetrical relation that indicates how strongly two individual propositions cohere. A connectionist computer program is then used to determine which hypothesis coheres best with the given observational evidence. By applying this method to consecutive states of a scientific controversy, Thagard seeks to explain its evolution.

Although we share with Thagard the aim to understand the dynamics of rational controversies, Thagard is primarily interested in explaining theory change in science (e.g., Why is it that some hypothesis was well corroborated at state t_1 but justified to a much lesser degree at a later state t_2 of the debate?), whereas the scope of our inquiry surely covers but is not restricted to scientific controversies. Moreover, by evaluating debates in terms of coherence, Thagard's model neither takes account of the agreement between proponent positions nor of their verisimilitude. As a consequence, it cannot assess the consensus- and truth-conduciveness of controversial argumentation, which is this inquiry's main mission. There are, of course, major differences concerning the specific representation of a debate as well. Most importantly, Thagard's account does not, unlike the theory of dialectical structures, represent inferential relations between the statements which figure in a debate (except for contradiction). His approach does thence not qualify as an argumentation-theoretic one in the first place. Yet, in spite of these basic theoretical differences, Thagard's method and the theory of dialectical structures yield seemingly similar results when applied to real debates. Specifically, Thagard's analyses of scientific controversies as explanatory maps evoke, immediately, argument maps that visualize dialectical structures. This superficial resemblance stems from the fact that an explanatory link, relating a couple of hypotheses on the one side with an observational item (the explanandum) on the other side, calls for an interpretation as argument (with the explanandum as conclusion), and a corresponding reconstruction according to the theory of dialectical structures. Hence, Thagard's concrete applications might actually be neatly transferred into the framework adopted throughout our inquiry.

Theories of judgement aggregation (cf. List and Puppe 2009; List and Polak 2010) study methods for merging various judgements (or sets of judgements, i.e., proponent positions) into a single, collective judgement (or a set of judgements, i.e., a proponent position). At the heart of this research program lies the observation that simple majority voting on individual sentences might aggregate consistent individual positions into an inconsistent collective one. So, consider three agents who assign truth values to the sentences $p \vee q$, $\neg q$, p as follows:

Agent	$p \vee q$	$\neg q$	p
1	True	True	True
2	True	False	False
3	False	True	False

Clearly, each agent holds a logically consistent position. Now, assume it as required to aggregate the mutually distinct positions into a collective judgement about the corresponding three sentences (e.g., because the agents belong to a jury in a U.S. court, or to a scientific advisory body, or to the Cabinet). A straightforward method for doing so is majority voting with respect to the individual statements. This yields, however,

	$p \vee q$	$\neg q$	p
Maj	True	True	False

which is an inconsistent truth-value assignment, hence the "discursive dilemma," as this problem is also referred to. Theories of judgement aggregation seek and study procedures for combining judgements which don't result in inconsistent collective judgements, provided the individual agents hold consistent positions. In analogy to Arrow's impossibility theorem for preference aggregation (Arrow 1963), List and Pettit (2002, 2004) have proven impossibility theorems for judgement aggregation, which demonstrate, generally, that such procedures cannot simultaneously meet a set of given, sought-after criteria.

Now, how do theories of judgement aggregation relate to the model of debate dynamics presumed in this inquiry? To begin with, theories of judgement aggregation neither assume agents to be logically omniscient nor to be logically ignorant. Instead, the discursive dilemma arises, and can be studied, based on the assumption of limited logico-argumentative capacities, which dovetails with the theory of dialectical structures. However, theories of judgement aggregation don't investigate how the beliefs of individual agents change given the introduction of new arguments or the discovery of new evidence. They presume, in contrast, that the rational debate has come to standstill, without having generated a universal consensus. The question addressed by theories of judgement aggregation reads: What should we do if (a) a consensus position has to be reached—for whatever reasons, if (b) the process of giving and taking reasons has come to an end because no new arguments or facts pertaining to the debate are discovered anymore, and if (c) a residual dissent persists nevertheless? This is of course an interesting and relevant question, yet it concerns a completely different phase of collective belief formation than the one studied in this book. Our investigation assesses the consensus- and truth-conduciveness of controversial argumentation, studying, in particular, whether proponent positions approach each other—and the truth—in the course of a debate, that is, by way of introducing new arguments. The point at which no new arguments are discovered, at which a controversy ends, delimits the scope of our inquiry. But it is precisely at this point where theories of judgement aggregation set in. So, a model of debate dynamics on the one hand and theories of judgement aggregation on the other hand, by virtue of relating to consecutive phases of social belief formation, rather complement, than compete with each other.

Chapter 2
An Introduction to the Theory of Dialectical Structures

2.1 Fundamental Concepts

A *dialectical structure* $\tau = \langle T, A, U \rangle$ is a set of deductive arguments (premiss–conclusion structure), T, on which an attack relation, A, and a support relation, U, are defined as follows ($a, b \in T$):

- $A(a,b) :\Longleftrightarrow$ a's conclusion is contradictory to one of b's premisses.
- $U(a,b) :\Longleftrightarrow$ a's conclusion is equivalent to one of b's premisses.[1]

This definition, as well as the entire following investigation, assumes that individual arguments are reconstructed as deductively valid—relative to some given but not further specified logical system. I shall, from time to time, refer to this system or set of inference rules as the reconstruction logic. In principle, the theory of dialectical structures may be combined with any reconstruction logic whatsoever.[2] Moreover, we shall assume that dialectical structures are set up with regard to a given pool of sentences S (consisting, e.g., of the sentences which are relevant

[1] A dialectical structures is a special type of bipolar argumentation framework as developed by Cayrol and Lagasquie-Schiex (2005). Cayrol and Lagasquie-Schiex extend the abstract approach of Dung (1995) by adding support relations to Dung's framework which originally considered attack relations between arguments only. A specific interpretation of Dung's abstract framework that analyzes arguments as premiss–conclusion structures is carried out in Bondarenko et al. (1997). The theory of dialectical structures is more thoroughly developed in Betz (2008, 2009), and in particular in Betz (2010).

[2] As we shall see later, the evaluation procedures provided by the theory of dialectical structures appeal, however, to some minimal logical principles such as the principle of noncontradiction. Still, as far as I can see, one may even consistently claim that this principle should not be relied on when reconstructing individual arguments (the reconstruction logic should not imply the law of noncontradiction) while maintaining that, for the evaluation of proponent positions in complex controversies, this principle may very well be assumed. The theory of dialectical structures is compatible with all sorts of reconstruction logics.

G. Betz, *Debate Dynamics: How Controversy Improves Our Beliefs*, Synthese Library 357, DOI 10.1007/978-94-007-4599-5_2,

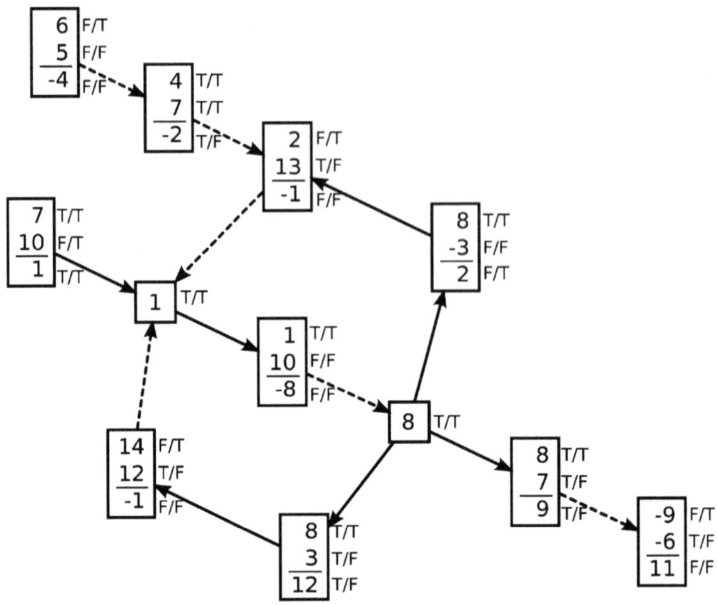

Fig. 2.1 A dialectical structure with two complete positions attached. Truth values are symbolized by "T" (true) and "F" (false)

in a given controversy). Premisses and conclusions of arguments in τ, this is, are members of S. We stipulate that this pool is closed with regard to negation ($p \in S$ implies $\neg p \in S$).

Complex debates can be reconstructed as dialectical structures.[3] Figure 2.1 depicts a purely formal example of a dialectical structure. Numbers stand for sentences, and a negative number denotes the negation of the sentence which is designated by the corresponding positive integer. Each box represents an argument or a thesis. Continuous and dashed arrows indicate the support and attack relationship, respectively.

Relative to a dialectical structure τ, which in a sense depicts the state of a debate, one can specify the positions of different proponents. We may, generally, distinguish complete and partial positions. A *complete position* $\mathscr{Q}$ (a proponent can adopt) on τ is a truth value assignment to all sentences in the relevant pool, that is, $\mathscr{Q} : S \rightarrow \{t, f\}$. A *partial position* $\mathscr{P}$ (a proponent can adopt) on τ is a truth value assignment to some sentences of that general pool, that is, $\mathscr{P} : S' \rightarrow \{t, f\}$, where $S' \subseteq S$. An *atomic position* (a proponent can adopt) on τ, finally, assigns exactly one sentence a truth value. As a handy notation, we refer to the position defined on $\{p_1, \dots, p_n\}$

[3]Cf. the online database on http://www.argunet.org

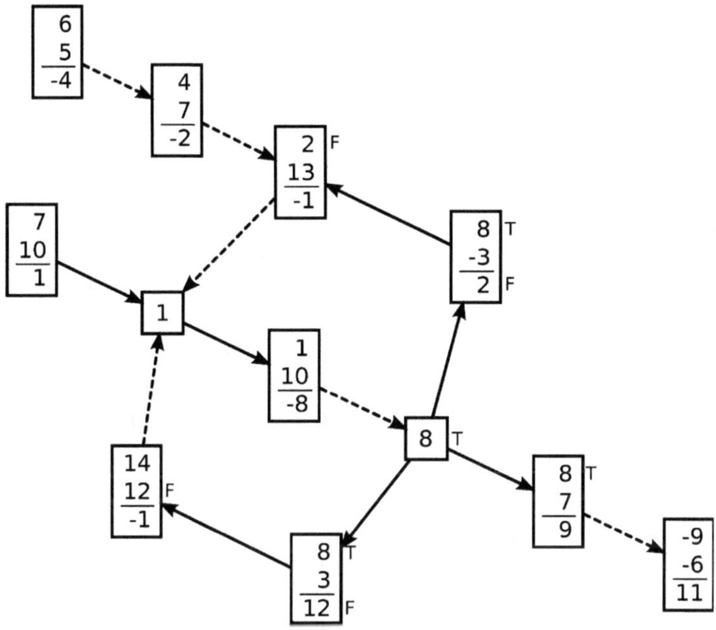

Fig. 2.2 A dialectical structure with one partial position attached

which assigns all sentences the value *true* by "$[p_1, \ldots, p_n]$." Whereas Fig. 2.1 shows two complete positions defined on a dialectical structure,[4] Fig. 2.2 gives an example for a partial position defined on the very same debate.

Partial positions can be combined. Let $\mathscr{P}_1 : S_1 \to \{t, f\}$ and $\mathscr{P}_2 : S_2 \to \{t, f\}$ be two partial positions which agree on $S_1 \cap S_2$. The *conjunction* of these positions, $(\mathscr{P}_1 \& \mathscr{P}_2) : S_1 \cup S_2 \to \{t, f\}$, can be defined by,

$$p \mapsto \begin{cases} \mathscr{P}_1(p) \text{ if } p \in S_1 \\ \mathscr{P}_2(p) \text{ if } p \in S_2 \setminus S_1 \end{cases}.$$

Obviously, the arguments that make up a dialectical structure impose certain constraints on which beliefs a proponent can reasonably maintain. Not every complete or partial position can be rationally adopted. Thus, a complete position $\mathscr{Q}$ on τ is *(dialectically) coherent* if and only if:

1. Contradictory sentences in S are assigned complementary truth values.
2. For every argument $a \in T$: if every premiss of a is, according to $\mathscr{Q}$, true, the conclusion is assigned the value *true*, too.

[4] With the pool of sentences being $\{-14, \ldots, -1, 1, \ldots, 14\}$, we tacitly assume that $-14, -13, -12, -11, -10, -7, -5$ are assigned truth values complementary to those assigned to $14, 13, \ldots$.

A partial position $\mathscr{P} : S' \to \{t, f\}$ on τ is *(dialectically) coherent* if and only if it can be extended to a complete position $\mathscr{Q}$ on τ ($\mathscr{P} = \mathscr{Q}|_{S'}$) which is coherent.

Returning to the complete positions depicted in Fig. 2.1, we may note:

- The left-hand-side complete position in that example is coherent. It complies with the coherence conditions set up above.
- The right-hand-side assignment of truth values does, however, not represent a coherent position. Over and above violating both coherence constraints, it isn't a well-defined position in the very first place. Note that, following the right-hand-side assignment, different tokens of sentence ten possess different truth values: Apparently, we haven't a function defined on S at all. Moreover, contradictory sentences $3/-3$ are both considered true, in conflict with the first coherence constraint. Finally, the right-hand-side truth value assignment violates the second constraint because the conclusion of argument $(4,7;-2)$ is false despite its premisses being true.

Furthermore, the partial position shown in Fig. 2.2 is not coherent, either, because it cannot be extended to a complete, coherent position. To see this, consider, first of all, the argument (8, 3; 12). Since premiss 8 is true and the conclusion 12 is false, any coherent position which extends the partial one has to declare the remaining premiss 3 as false. Otherwise, the complete position would violate the second coherence constraint. The same reasoning applies to argument $(8,-3;2)$: Sentence -3 has to be false for analogous reasons. Consequently, -3 and 3 would possess the same truth value, which contradicts the first coherence constraint. Hence, the partial position cannot be extended to a coherent complete position.

Background knowledge, with regard to which proponent positions are evaluated, may be represented as a partial position $\mathscr{B}$ on τ. A position is *coherent relative to some background knowledge* $\mathscr{B}$ if and only if it is (1) dialectically coherent and (2) extends $\mathscr{B}$. Evaluating a debate against some background knowledge hence merely implies to diminish the set of coherent positions one takes into account.

It is helpful, in order to reduce the complexity of the following analysis, to assume that positions are declared on half of the sentences belonging to S, only. More specifically, a position shall assign a truth value to exactly one sentence each of every pair of contradictory sentences belonging to S. This brings down the dimension of the boolean vectors by a half: Every position assigns $n = |S|/2$ truth values. We shall assume that the truth values of the remaining sentences in S are determined so that the first coherence constraint is satisfied.

2.2 Degrees of Justification

Based on the primitive notions put forward in the preceeding section, we can now introduce the concepts of dialectic entailment and degree of partial entailment. Thus, a partial position $\mathscr{P}_2$ *dialectically entails* a partial position $\mathscr{P}_1$ if and only if all coherent and complete positions which extend $\mathscr{P}_2$ equally extend $\mathscr{P}_1$. In the example above (Fig. 2.1), the partial position according to which 8 and 10 are true dialectically entails the atomic positions which assign 1 the value *false*.

The concept of dialectic entailment may be generalized by following Wittgenstein's basic idea in the *Tractatus* (and identifying "cases" with "complete and coherent positions" on τ): The *degree of partial entailment* of a partial position $\mathscr{P}_1$ by a partial position $\mathscr{P}_2$ can be defined as

$$\begin{aligned}\textsc{Doj}(\mathscr{P}_1|\mathscr{P}_2) &:= \frac{\text{number of cases with } \mathscr{P}_1 \text{ \& } \mathscr{P}_2}{\text{number of cases with } \mathscr{P}_2} \\ &= \frac{\begin{array}{c}\text{number of complete \& coherent positions}\\ \text{that extend } \mathscr{P}_1 \text{ \& } \mathscr{P}_2\end{array}}{\begin{array}{c}\text{number of complete \& coherent positions}\\ \text{that extend } \mathscr{P}_2\end{array}} \\ &= \frac{\sigma_\tau(\mathscr{P}_1,\mathscr{P}_2)}{\sigma_\tau(\mathscr{P}_2)}, \end{aligned} \tag{2.1}$$

with $\sigma_\tau(\mathscr{P},\mathscr{Q},\ldots)$ denoting the number of complete and coherent positions on τ which extend every position $\mathscr{P},\mathscr{Q},\ldots$. As a consequence, $\textsc{Doj}(\mathscr{P}_1|\mathscr{P}_2) = 1$ if and only if $\mathscr{P}_2$ dialectically entails $\mathscr{P}_1$. Degrees of partial entailment, thus defined, satisfy the *axioms of probability theory*.[5]

Reconsider the example depicted in Fig. 2.1. Whereas 8 and 10 dialectically entail -1, as noted above, 8 alone does not. Yet, 8 entails -1 to some extent, and we are now in a position to quantify this degree of partial entailment. Namely,

$$\textsc{Doj}([-1]|[8]) = \frac{\sigma_\tau([-1],[8])}{\sigma_\tau([8])} = \frac{230}{281} \approx 0.82.$$

Compare this with the modest degree of partial entailment for the comparatively distant and unrelated sentences 8 and 5:

[5]See Betz (2011b). Note also that degrees of partial entailment are defined with regard to partial positions and not with regard to sentences. The corresponding measure on the set of sentences—$\mathrm{P}(p) := \textsc{Doj}([p])$—does not necessarily satisfy the Kolmogoroff axioms. This is the problem: For every probability measure over a set of statements, it holds that $\mathrm{P}(p \vee q) = \mathrm{P}(p) + \mathrm{P}(q)$ for contrary p, q. Now, assume that the three sentences $p \vee q$, p, and q figure in some τ and that there is no dialectically coherent position according to which both p and q are true. Still, this does not guarantee that the (unconditional) degrees of partial entailment of the atomic positions according to which p and, respectively, q are true, add up to the (unconditional) degree of partial entailment of the atomic position which says that $p \vee q$ is true. This is because not every coherent complete position according to which p is true assigns $p \vee q$ the value *true*—unless an argument like $(p;p \vee q)$ is included in τ. A similar reasoning applies to conjuncts of single statements. Thus, degrees of partial entailment, when defined on sentences and not on partial positions, satisfy the probability axioms only if the respective dialectical structure is suitably augmented by simple arguments as indicated.

$$\textsc{Doj}([5]|[8]) = \frac{\sigma_\tau([5],[8])}{\sigma_\tau([8])} = \frac{161}{281} \approx 0.57.$$

Finally, the *degree of justification* of a partial position $\mathscr{P}$, or, as we shall say alternatively, its *robustness*, can be defined as its degree of partial entailment from the empty set:

$$\textsc{Doj}(\mathscr{P}) := \textsc{Doj}(\mathscr{P}|\emptyset) = \frac{\text{number of complete \& coherent positions that extend } \mathscr{P}}{\text{number of complete \& coherent positions}}. \tag{2.2}$$

It can be shown that degrees of justification possess the following properties (cf. Betz 2011b):

- Introducing an independent argument that supports (attacks) some thesis t increases (decreases) t's degree of justification.
- Introducing an independent argument that supports (attacks) a supporting argument for some thesis t increases (decreases) t's degree of justification.
- Introducing an independent argument that supports (attacks) some argument which attacks thesis t decreases (increases) t's degree of justification.

Consider the degrees of justification of the sentences 1 and 8 in our standard example (more precisely, the degrees of justification of the atomic positions according to which 1, respectively 8, is true):

$$\textsc{Doj}([1]) \approx 0.37,$$
$$\textsc{Doj}([8]) \approx 0.23.$$

Why is the unconditional degree of justification of 8 much lower than of 1? We can understand this by noting, first of all, that every argument with a premiss p may also be reconstructed as a counterargument against p. The very inferential relation encoded in (8, 7; 9) is equally expressed by the argument $(-9,7;-8)$. This said, thesis 8 in our standard example is virtually attacked by four arguments, without receiving any support. Thesis 1, however, is merely attacked by three arguments and supported by one. This explains why $\textsc{Doj}([1]) > \textsc{Doj}([8])$.

Sentences which possess maximally high or low degrees of justification are special. A sentence p is:

- τ-true iff p is true in all coherent complete positions, that is, $\textsc{Doj}([p]) = 1$.
- τ-false iff p is false in all coherent complete positions, that is, $\textsc{Doj}([p]) = 0$.
- τ-analytic iff p is τ-true or -false.

As a dialectically incoherent position is, relative to the reconstruction logic, logically inconsistent, every sentence which is τ-true or -false is a logical tautology or a logical falsehood.

2.3 The Space of Coherent Positions

Let τ be some dialectical structure. The argument map corresponding to τ represents the space of reasons with regard to which proponent positions can be located, and evaluated. We shall, in this section, introduce the complementary concept of the *space of coherent positions* corresponding to τ. Let Γ_τ be the set of all coherent (complete) positions on τ. This set and its internal structure make up the space of coherent positions.

First and foremost, a metric on Γ_τ can be introduced straightforwardly. Thus, the *normalized distance* between two complete positions $\mathscr{P}, \mathscr{Q}$ is defined as their Hamming distance divided by the number of pairs of contradictory sentences on which the positions are declared ($n = |S|/2$), that is,

$$\Delta(\mathscr{P}, \mathscr{Q}) := \frac{\mathrm{HD}(\mathscr{P}, \mathscr{Q})}{n}.$$

Obviously, $0 \leq \Delta(\mathscr{P}, \mathscr{Q}) \leq 1$ for any two coherent positions. We have, moreover, with n^* denoting the number of τ-analytic sentences in τ, $1 - \Delta(\mathscr{P}, \mathscr{Q}) \geq \frac{n^*}{n}$, since any two coherent positions agree at least with regard to the τ-analytic sentences.

The normalized distance Δ can be used to define *normalized agreement* between two positions, $\mathscr{P}, \mathscr{Q}$, namely, as $1 - \Delta(\mathscr{P}, \mathscr{Q})$. If, moreover, $\mathscr{T}$ is the objectively correct assignment of truth values, that is, the true position, we refer to the normalized agreement of some position with $\mathscr{T}$ as its *verisimilitude*. Note that this amounts to a very simple and unambitious, syntactic explication of verisimilitude. Basically, the degree of truthlikeness of a complete or partial position is determined by counting the number of correct truth value assignments it comprises. As a consequence, atomic positions, that is, individual beliefs, possess either a verisimilitude of 1 or 0. Thus, our definition deviates importantly from Popper's concept of verisimilitude (Popper 1963, p. 316), which runs into various problems, and similar concepts that try to capture the degree of truthlikeness of individual statements (see, for reviews, Niiniluoto 1998; Oddie 2008).[6] Despite its simplicity, the concept of verisimilitude, as introduced above, relates to various analyses in the literature. To start with, our notion of verisimilitude translates seamlessly into Goldman's concept

[6]Moreover, we effectively avoid counterexamples of the kind "The partial position ⟨'Our solar system contains 12 planets,' 'Helium is lighter than air'⟩ actually seems to be much closer to the truth than the partial position ⟨'Our solar system contains 12,000 planets,' 'Helium is lighter than air'⟩, yet both positions exhibit a verisimilitude of 0.5" by comparing positions with regard to their verisimilitude only if they range over one and the same set of sentences. Now, these advantages of our very simple notion of verisimilitude, however, go hand in hand with limitations and shortcomings. In particular, the straightforward and unambitious concept of verisimilitude, no matter how useful it might turn out to be in the following investigation, does not capture every aspect of our everyday concept of truthlikeness. So, the strong intuition that, for example, "Earth is 3 billion years old" is much closer to the truth than "Earth is 700,000 years old," is not, at least not directly, accounted for.

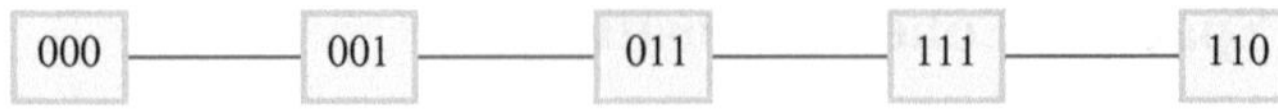

Fig. 2.3 A space of five coherent positions visualized as a graph. The IO-strings represent truth values assigned to three sentences in the pool, assuming that the positions agree with regard to the other sentences. The distance between 001 and 110 in the graph equals 4, whereas the Hamming distance between the two positions is 2

of veritistic value (cf. Goldman 1999, p. 89). As long as we consider proponents who hold complete positions and thence don't withhold judgement, a position's verisimilitude is nothing but the mean veritistic value of the proponent's doxastic states. Concerning the verisimilitude debate in the wake of Popper (1963), some explications of the comparative notion of truthlikeness, which have been suggested, cohere remarkably with the concept of verisimilitude as introduced above.[7]

Now, how can individual agreement and verisimilitude values be used to characterize an entire state of a debate? Let $\mathcal{P}^1, \ldots, \mathcal{P}^m$ be the proponent positions in a debate at some step t. We define the debate-wide mean normalized agreement at step t as the average of all pairwise normalized agreements, that is, as

$$\binom{m}{2}^{-1} \sum_{i=2}^{m} \sum_{j=1}^{i-1} \left(1 - \Delta(\mathcal{P}^i, \mathcal{P}^j)\right).$$

In addition, the debate-wide mean verisimilitude is simply defined as the average of the proponent positions' individual verisimilitude values.

The entire space of coherent positions can be visualized as an undirected graph. In such a visualization, every complete and coherent position is represented as a vertex. Two vertices are connected by an edge if and only if the Hamming distance between the corresponding positions equals one (i.e., is minimal). Such a kind of representation of the space of coherent positions, though, has to be read with care. Most importantly, the distance between two vertices in the graph is not necessarily equal to the Hamming distance between the corresponding positions (see Fig. 2.3 for a counterexample). Taking this into account, such graph-theoretical representations

[7]To see this in some more detail, note that the concept of a constituent, as used, for example, by Kuipers and Schurz (2011) and Niiniluoto (2011), corresponds to our notion of a complete position. As Kuipers and Schurz (2011) remark, the Hamming distance between constituents is fundamental for defining verisimilitude (see also Riegler and Douven 2009; de Lavalette and Zwart 2011). With respect to the so-called BF-approach developed by Cevolani, Crupi, and Festa (e.g., Cevolani et al. 2011), we may, more specifically, identify the verisimilitude of a complete position $\mathcal{Q}$ with its "degree of true b-content"; the verisimilitude of a partial position $\mathcal{P}$ (as defined here) is, moreover, proportional to the "degree of true b-content" of $\mathcal{P}$ (understood as a "conjunctive theory" in line with Cevolani et al. 2011)—precisely, it equals n/m times its "degree of true b-content." Finally, a partial position $\mathcal{P}_2$ displays greater verisimilitude than a partial position $\mathcal{P}_1$ in terms of our framework, if $\mathcal{P}_2$ is "more verisimilar" than $\mathcal{P}_1$ according to the definition proposed by Cevolani et al. (2011). The reverse, it seems however, does not hold in general.

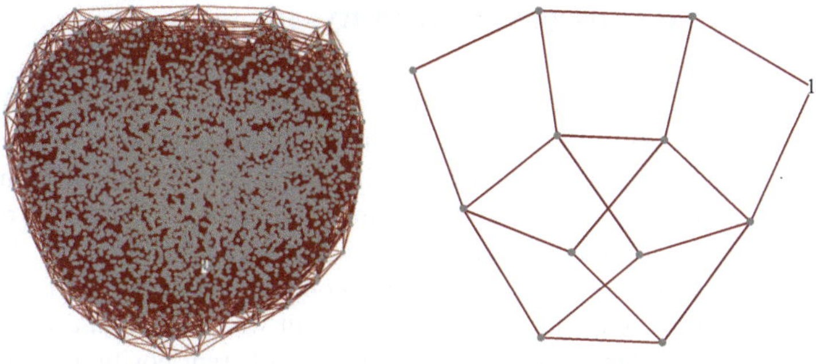

Fig. 2.4 The space of coherent positions corresponding to the dialectical structure depicted in Fig. 2.1. The *left-hand panel* shows the entire space of coherent positions; the *right-hand panel* displays merely a section of that space, namely with regard to sentences 2, 3, 8, and 12

may nevertheless convey an approximately correct idea of the geometry of the space of coherent positions and will prove useful in the following chapters' analyses.

Plotting the entire space of coherent positions for sufficiently large debates, however, is illusory. The number of vertices grows exponentially when increasing the pool of sentences; soon, individual edges can hardly be identified. A convenient remedy is to plot but a lower dimensional section of the entire space of coherent positions, or, in other words, to plot the space of coherent partial positions declared on a strict subset of S. Nonidentical complete positions may collapse onto one and the same partial position when being projected on a lower dimensional subspace (different complete positions may extend one and the same partial position). This typically reduces the number of coherent positions which have to be plotted significantly. What does such a lower dimensional plot tell us about the geometry of the entire space of coherent positions? If two nodes (representing partial positions) aren't connected in a section plot, they disagree at least with regard to two sentences belonging to the corresponding subset of S and thus disagree at least with regard to two sentences in S. As a consequence, two coherent positions whose lower dimensional projections aren't linked in a sectional plot, aren't connected in a higher dimensional plot, either. Since every coherent complete position is mapped onto some position in the lower dimensional plot, studying the latter allows one to identify clusters of positions that remain isolated when reincluding further sentences (dimensions).

Figure 2.4 plots the space of coherent positions of our standard example. The left-hand panel shows the entire (14-dimensional) space of coherent positions. This results in 4,876 different vertices. The right-hand graph, in contrast, plots a 4-dimensional section of that very space by displaying all coherent partial positions which are declared on the sentences 2, 3, 8, and 12.

2.4 Normalized Closeness Centrality

The geometry of the space of coherent positions, as well as its dynamic deformation in the course of a debate, will be crucial for understanding the position dynamics in controversial argumentation. Although plotting lower dimensional sections allows one, at least, to generate partial visualizations of the space of coherent positions, this method faces clear limitations and is primarily suited for heuristic or illustrative purposes. An accurate analysis of the space of coherent positions, and its geometry, is in need of appropriate quantitative measures. But what exactly are we supposed to measure? An important feature of the space of coherent positions is its overall compactness: Are all coherent positions closely related to each other or is the space rather stretched out and detached, the distances between coherent positions being relatively large? A similar property, relating to single positions, consists in whether a coherent position is situated in a compact space, with close relations to the other positions, or whether it is rather detached from the other positions. This last property of a single position depends both on the geometry of the entire space of coherent positions as well as on the specific location the position occupies within that very space.

In order to make this idea of a position being closely related to other positions more precise, we borrow and adapt the graph-theoretical concept of closeness centrality. Normalized closeness centrality of some position $\mathscr{P}$ relative to a set of positions $\mathbf{A}$, $\mathscr{P} \in \mathbf{A}$, measures how centrally $\mathscr{P}$ is located in the set of positions. We define it as the inverse of the mean normalized distance between $\mathscr{P}$ and the positions in $\mathbf{A}$, divided by 2:

$$\begin{aligned} \text{NCC}(\mathscr{P},\mathbf{A}) &:= \frac{1}{2} \cdot \frac{1}{(\Sigma_{\mathscr{Q}\in\mathbf{A}} \Delta(\mathscr{P},\mathscr{Q}))/|\mathbf{A}|} \\ &= \frac{|\mathbf{A}|}{2\Sigma_{\mathscr{Q}\in\mathbf{A}} \Delta(\mathscr{P},\mathscr{Q})}. \end{aligned} \tag{2.3}$$

$\text{NCC}(\mathscr{P},\mathbf{A})$ is controlled both by the geometry of $\mathbf{A}$ as well as by the specific location $\mathscr{P}$ occupies in $\mathbf{A}$: For a given $\mathbf{A}$, NCC depends on whether $\mathscr{P}$ is centrally (high NCC) or remotely (low NCC) situated. Whether there exist highly remote or highly central parts of $\mathbf{A}$ at all (which may be occupied by $\mathscr{P}$) is obviously determined by $\mathbf{A}$'s geometry. If all positions in $\mathbf{A}$ possess a relatively high (low) NCC—with regard to $\mathbf{A}$—this means that $\mathbf{A}$ is rather compact (dispersed). Consider Fig. 2.5 for some illustrative NCCs.

NCC is normalized so as to ensure that in a space of coherent positions which consists in all combinatorially possible truth value assignments on some pool of sentences, every position's NCC equals 1. In other words: In the absence of any arguments whatsoever, every positions exhibits a normalized closeness centrality of 1. This is what the following theorem verifies. Note that the space of coherent

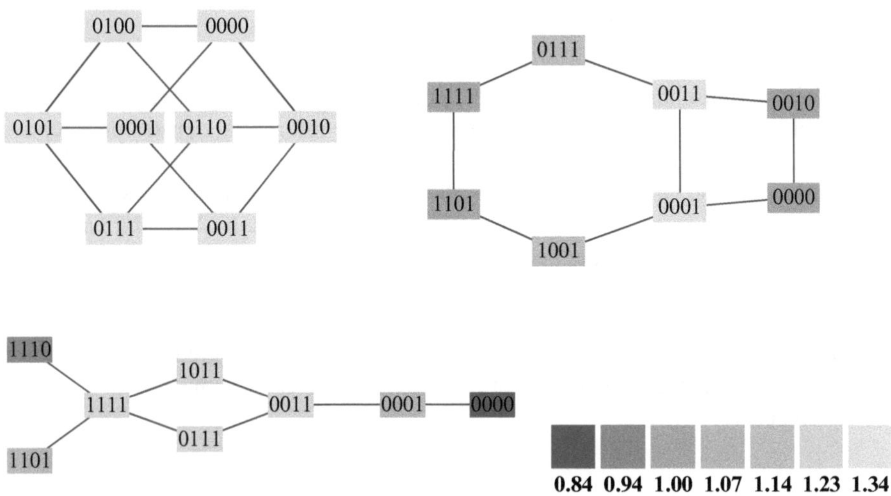

Fig. 2.5 Representations of three spaces of coherent positions declared on four sentences each. Vertex shading is a function of the respective normalized closeness centrality. Central positions exhibit greater NCC than remote ones. In addition, average NCC on compact spaces is, compared to those on stretched, detached, and disconnected spaces, higher

positions on a pool S is, in the absence of arguments, isomorphic to the n-dimensional hypercube—each proponent position corresponding to exactly one of its vertices.

Proposition 1 (NCC in a hypercube). *Let* **H** *be the set of all positions given a pool of sentences S of size $2n$. Every position $\mathscr{P}$ in* **H** *possesses a normalized closeness centrality of 1 relative to* **H**.

Proof. Let $\mathscr{P}$ be some position in **H**. There are $\binom{n}{i}$ positions in **H** whose Hamming distance to $\mathscr{P}$ is i. We thus have, according to (2.3),

$$\begin{aligned} \mathrm{NCC}(\mathscr{P},\mathbf{H}) &:= \frac{|\mathbf{H}|}{2\sum_{\mathscr{Q}\in\mathbf{H}}\Delta(\mathscr{P},\mathscr{Q})} \\ &= \frac{2^n}{2\sum_{\mathscr{Q}\in\mathbf{A}}\mathrm{HD}(\mathscr{P},\mathscr{Q})/n} \\ &= \frac{n2^n}{2\sum_{i=0}^{n}\binom{n}{i}i} \\ &= \frac{n2^n}{2\times 2^{n-1}n} = 1. \end{aligned}$$

□

2.5 Inferential Density

A further pivotal characteristic of a dialectical structure is its inferential density. Intuitively, this can be understood as measure of the inferential constraints encoded in τ. Roughly, the more arguments a dialectical structure hosts, the higher its inferential density. However, not every additional argument changes the inferential relations encoded in τ—some arguments are redundant and don't render any previously coherent position dialectically incoherent. It is thus appropriate to explicate the notion of a dialectical structure's inferential density in terms its corresponding space of coherent positions rather than in terms of the argument map itself. The smaller the number of coherent positions (left) on some τ, the greater its inferential density. More precisely, we define the inferential density of some τ with a pool of $2n$ sentences as

$$D(\tau) := \frac{n - \lg(\sigma_\tau)}{n}, \tag{2.4}$$

where $\sigma_\tau = |\Gamma_\tau|$ refers to the number of coherent and complete positions on τ. The inferential density, thus defined, relates the number of binary choices that one has to make when adopting a position in τ to the number of (initial) binary choices which determine a position in the absence of any arguments. Whatever the arguments contained in τ, n binary choices (one true/false choice for every pair of contradictory sentences) are necessary to specify a complete position. Yet, it requires but $\lg(\sigma_\tau)$ binary choices to pick one of the σ_τ coherent positions. Obviously, $D(\tau)$ is strictly monotonic in σ_τ: The fewer coherent positions, the higher the inferential density. Moreover, since $\sigma_\tau \leq 2^n$, we have

$$D(\tau) = \frac{n - \lg(\sigma_\tau)}{n} \geq \frac{n - \lg(2^n)}{n} = 0.$$

Let's consider some extreme cases. If there is no coherent position on τ at all, $\sigma_\tau = 0$, we have $D(\tau) = \infty$. In this case, the reconstruction logic (the inference rules underlying the individual arguments) is inconsistent. If exactly one complete position is coherent, $\sigma_\tau = 1$, then $D(\tau) = 1$ and all sentences are τ-analytic. If $\sigma_\tau = 2$, we have $D(\tau) = (n-1)/n$. Finally, if every combinatorically possible position is coherent, $\sigma_\tau = 2^n$ and $D(\tau) = 0$; the dialectical structure imposes no inferential constraints whatsoever.

The inferential density of our standard example depicted in Fig. 2.1 equals 0.30. Figure 2.6 displays the dialectical structure of a real argumentation—Descartes' Meditations as reconstructed in Betz (2011a). The reconstruction consists in 78 arguments which draw their premisses and conclusions from a pool of 2×280

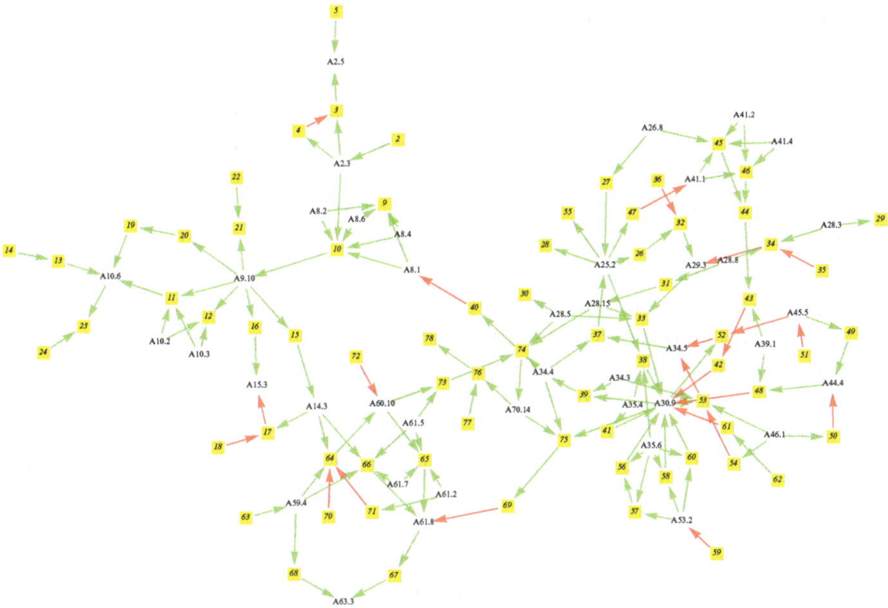

Fig. 2.6 The dialectical structure of Descartes' Meditations according to the reconstruction in Betz (2011a). The argument map depicts the arguments (*yellow boxes*) and the pivotal theses of the argumentation (*plain labels*)

sentences.[8] An argument possesses, on average, 4 premisses. The inferential density of the Meditations, accordingly reconstructed, equals 0.02 and is, thus, significantly lower than the standard example's one.

The case of Descartes' Meditations raises the question which densities one may attain in real debates at all. Given that we will simulate debates up to very high densities (typically 1) in the subsequent chapters, this becomes even more important a question. For what does a debate simulation tell us about real argumentation if it operates at densities which are, in real controversies, never reached in the first place? One of our first, yet crucial, stipulations requires that individual arguments be deductively valid: If all the premisses are true, the conclusion is necessarily true, as well. So, when reconstructing a real debate, only universally shared inference rules must be presumed. Otherwise, the inferences in the reconstructed arguments might simply not be truth preserving. In the reconstruction of the Meditations, only a small subset of the inference rules systematized by classical logic are used. But founding a reconstruction on a minimal base of inference rules multiplies, in the same time, the number of implicit assumptions which have to be made explicit as

[8]Five of the arguments are not related to the main argumentation and thus omitted in the graph. Moreover, some of the arguments as reconstructed in Betz (2011a) are split up into several parts so as to make the dialectical role of preliminary conclusions explicit.

additional premisses. And clearly, the more premisses, the lower, *ceteris paribus*, the inferential density. This suggests that the inferential density of a debate depends on a fundamental choice which is made in the course of its reconstruction, that is, which inference rules, general principles, and further statements to consider as universal background knowledge of the debate and thus not to include explicitly as additional premisses in the arguments.[9] The more generous and broader the body of universal background principles, the less premisses per argument and, therefore, the higher the debate's inferential density. As previously noted, the Descartes reconstruction is based on a small set of inference rules. Just to illustrate the effect of widening the body of background principles, assume that 50% of the premisses can be considered as universally shared, for example, as incontestable analytic truths. Removing half of the premisses (randomly chosen) from the debate's arguments increases its inferential density to 0.3–0.4. The exact increase obviously depends on which sentences—pivotal or remote ones—are removed. If, however, even 80% of the sentences figuring in the reconstruction may count as universally shared, the density rises up to 0.7. Thus, to the extent that we can reasonably assume, when interpreting a debate, a body of universally shared principles and facts, real controversies may well attain very high densities, even densities close to 1. Without such a body of background beliefs, however, the reconstruction has to stick to a minimal set of inference rules which are truth preserving by virtue of the meaning of so-called logical constants and which are shared by any competent speaker of our language. In this case, densities are typically relatively low. In sum, introducing background knowledge and omitting the premisses which are part of it results in debates displaying higher densities. Simulations of debates with high densities might tell us something about these cases. Removing background premisses from a debate's reconstruction altogether is rather primitive a way for dealing with background beliefs. Instead of eliminating background principles, the corresponding sentences might simply be assigned specific truth values, as indicated in Sect. 2.1. This represents a more sophisticated way of representing background knowledge—background knowledge doesn't shape a debate's reconstruction but is specified once the reconstruction is complete—and does, apparently, not affect a debate's inferential density. The explicit representation of background knowledge is clearly preferable when studying, for example, the effects of modifications of the background knowledge. In Chaps. 5 and 12, we investigate how explicitly incorporated background knowledge changes the consensus and veritistic dynamics of controversial debates, respectively.

In the remainder of this section, we study how to approximate inferential density and try to relate this discussion to the question which levels of inferential density can really be attained.

The following calculation derives a function that describes, at least as a rough approximation, the evolution of $D(\tau)$ as a function of the number of arguments put forward. To start with, we assume that new arguments are introduced successively

[9] See also Sect. 1.6.

into the debate, one at each step t, with τ initially containing no arguments ($A_0 = \emptyset$). Every argument consists of k premisses. What happens to the space of coherent positions if such an additional argument is introduced? We consider the $k+1$ sentences the new argument contains. There are 2^{k+1} partial positions on these sentences. At least for small inferential densities, these positions are coherent. We shall assume, and that is the first approximation, that irrespective of the inferential density and the number of arguments put forward, all of these partial positions are indeed coherent. The newly introduced argument, then, eliminates exactly one of these 2^{k+1} partial positions. As a second approximation, we presume that every partial position is extended by the same number of complete and coherent positions on τ. Putting forward the argument reduces σ_τ, consequently, by a factor $\frac{2^{k+1}-1}{2^{k+1}}$. Due to the introduction of ever new arguments, the space of coherent positions undergoes exponential decay, described by the following function:

$$\begin{aligned}\sigma_{\tau_t} &= 2^n \cdot \left(\frac{2^{k+1}-1}{2^{k+1}}\right)^t \\ &= 2^n \cdot (1-2^{-1-k})^t.\end{aligned}$$

Plugging this into the definition of inferential density (2.4) gives,

$$\begin{aligned}D(\tau_t) &= \frac{n - \lg\left(2^n \cdot (1-2^{-1-k})^t\right)}{n} \\ &= \underbrace{-\lg\left(1-2^{-1-k}\right)n^{-1}}_{\text{const.}}\, t. \qquad (2.5)\end{aligned}$$

Inferential density is, at least approximately, a linear function of the number of arguments in τ. Figure 2.7 displays how the density of debates increases with the number of arguments (t), and how this increase, in turn, depends on the overall size of the sentence pool (n) and the number of premisses each argument contains (k). The smaller the sentence pool and the smaller the size of each argument, the steeper the increase in inferential density.

Moreover, the linear approximation of inferential density (2.5) not only says that density is linearly correlated with the size of the debate, it also tells us, in effect, that it is the ratio of arguments per sentences, t/n, which crucially determines inferential density. Debates with the same number of premisses per argument, and the same number of arguments per sentence, possess, approximately, the same inferential density—regardless of the actual number of arguments in τ.

According to the approximation (2.5), a debate in which every sentence of the sentence pool is, on average, supported or attacked by exactly one argument ($t/n = 1$) possesses an inferential density of 0.19 (0.09, 0.04, 0.02) if its arguments contain 2 (3, 4, 5) premisses each. In order to attain an inferential density of 0.5, or higher, a debate with 2 (3, 4, 5) premisses per argument would have to hold, on average, 2.6 (5.4, 11.0, 22.0) arguments *per sentence* in S. Given that, in real debates, there

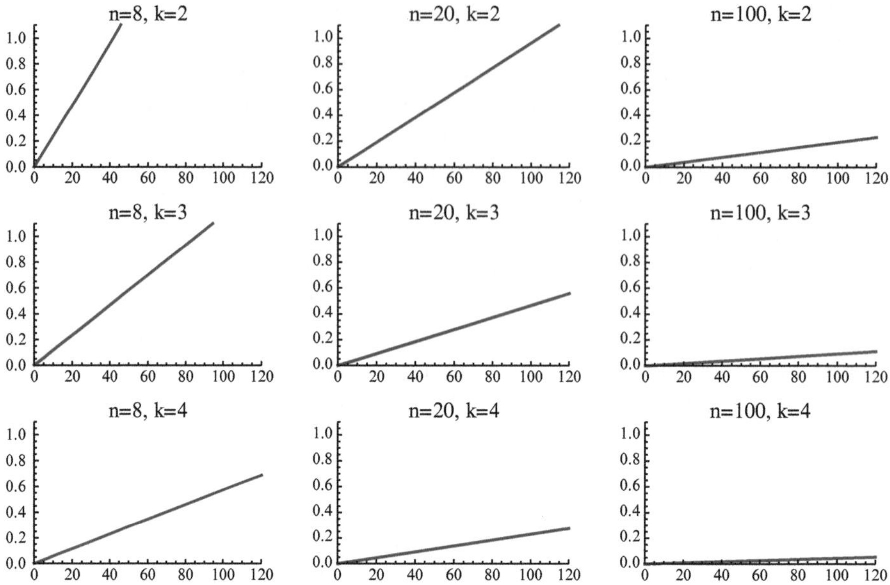

Fig. 2.7 Linear approximations of inferential density evolutions as a function of the number of arguments in τ (t). The parameters which control the rate of the increase are premisses per argument (k) and size of the entire sentence pool (n)

are typically far less arguments than sentences, and many sentences indeed remain unsupported and unattacked, the approximation suggests that natural argumentation hardly displays densities greater than 0.5.

But how accurate and reliable is our linear approximation (2.5) at all? The standard example (Fig. 2.1) with $t = 10, n = 14, k = 2$ should possess, according to the approximation, a density of 0.13, only. This is far-off from the actual 0.3. Likewise, the approximated density for the Meditations ($t = 78, n = 280, k = 4$) is 0.012 as compared to the real value of 0.021. The assumptions we made when deriving the approximation, it turns out, underestimate the effect of additional arguments on the number of coherent positions in τ.

While the examples suggest that, by relying on several approximations, we have significantly underestimated the effect of introducing new arguments on the inferential density of a debate, Fig. 2.8 shows that the linear approximation derived above might not be that deficient, at least not for all sorts of debates, in the end. This figure compares simulated mean inferential density evolutions on the one hand with our linear approximation on the other hand. The mean density evolutions are derived from the ensembles of debate simulations described in Chaps. 4 and 7. At least for randomly evolving debates and for densities below 0.8, the linear function (2.5) is not that bad an approximation, underestimating the effect of additional arguments only slightly. Debates with more sophisticated argumentation mechanisms (*multiple convert*, *multiple undercut*), however, possess considerably higher densities than predicted by Eq. (2.5).

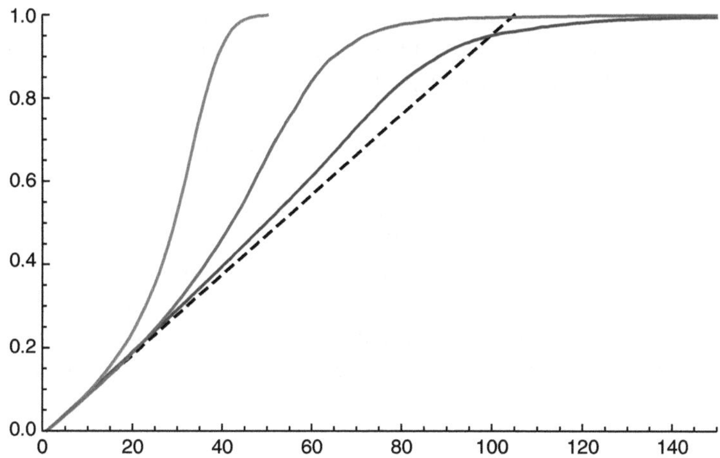

Fig. 2.8 Comparison of the linear approximation of inferential density (*dashed*) with the mean inferential density as a function of t derived from the ensemble simulations presented in Chaps. 4 and 7, that is, debates with *random argumentation* (*dark shading*, *right*), *multiple convert* argumentation (*medium shading*, *middle*), and *multiple undercut* argumentation (*light shading*, *left*). The simulated debates in these ensembles are based on a pool of 2×20 sentences and consist of arguments with 2 premisses each. The parameters of the linear approximation are, correspondingly, $n = 20$ and $k = 2$

2.6 The General Design of the Simulations

All simulations of debate dynamics presented in this report exhibit the same general design. Given a fixed pool of $2 \cdot 20$ sentences, the evolution of the dialectical structure and the evolution of a certain number of proponent positions are simulated in discrete time steps. More specifically, for each time step $t = 1, 2, 3, \ldots$ the dialectical structure τ_t as well as complete proponent positions $\mathcal{P}_t^1, \mathcal{P}_t^2, \ldots, \mathcal{P}_t^m$ are calculated. In the initial state of such a simulation, no arguments have been put forward ($\tau_0 = \langle \emptyset, \emptyset, \emptyset \rangle$). In case the veritistic dynamics are simulated, a randomly chosen truth value assignment $\mathcal{T}$ is marked as the objectively correct assignment of truth values. The proponents' initial positions are randomly determined, as well.[10] Three mechanisms drive the dynamics: (1) The *argumentation mechanism* determines τ_{t+1}

[10]More specifically, the way initial proponent positions are assigned depends on whether the simulation serves to study consensus- or truth-conduciveness, i.e. whether it is presented in this report's first or second part. In simulations of *consensus dynamics*, half of the initial positions consist in randomly and independently determined truth values, which are assigned to the individual sentences in the pool. For each position $\mathcal{P}$ specified in this way, an additional, corresponding proponent position is set up by (1) choosing a random number j with $0 \leq j \leq n$ and (2) altering j different truth value assignments of $\mathcal{P}$. This procedure has the effect that the relative frequency of extreme distances between proponents positions—compared to a random determination of proponent positions without adjustment (i.e. every initial position is obtained by independent random assignments of truth values)—is relatively high, as the following figure illustrates.

given τ_t and possibly further information; (2) the *discovery mechanism* describes the background knowledge $\mathcal{B}_t$, relative to which proponent positions are evaluated; (3) the *update mechanism* specifies how positions are updated, i.e. determines $\mathcal{P}^i_{t+1}$ (for every $i = 1 \ldots m$) as a function of τ_{t+1}, $\mathcal{B}_{t+1}$, $\mathcal{P}_{ti}$, and possibly further information.

In the simulations of veritistic debate dynamics, presented and discussed in Part II, the randomly chosen true position, $\mathcal{T}$, enables one not only to measure the normalized agreement of proponent positions with the truth, i.e. their verisimilitude, but is also used to check, semantically, newly introduced arguments for deductive validity. Such a semantic check belongs to every *argumentation mechanism* employed in Part II: As a deductive argument cannot possess true premisses and a false conclusion, we require that no argument introduced into a debate simulation renders the true position dialectically incoherent. Clearly, not to possess true

---

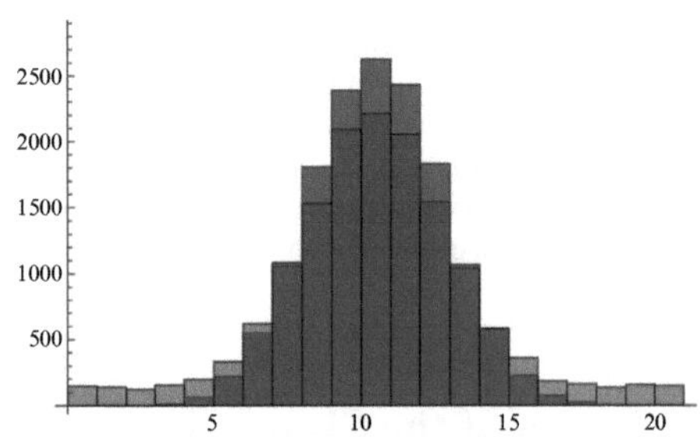

This figure plots the absolute frequencies of pairwise Hamming distances between initial proponent positions in an ensemble that consists of 1,000 debates with 6 positions each. Dark bars depict the frequencies without adjustment, light bars show the frequencies obtained with an adjusted random assignment of initial propositions. Due to the adjustment, the frequencies of position pairs with medium distance decrease, whereas those of positions with extreme distances increase.

The assignment of initial positions in simulations of *verisimilitude dynamics* deviates from the procedure just outlined only with respect to the very first step. Thus, initial positions of half of the proponents are constructed as fellows: The eventual verisimilitude of the initial position is fixed (with all verisimilitude values being equally likely), before a truth value assignment, which exhibits the corresponding verisimilitude, is chosen randomly. The remaining initial positions are specified as described above. This particular sampling method results in a more even initial distribution of (a) pairwise distances as well as (b) verisimilitude values, as the two following diagrams illustrate.

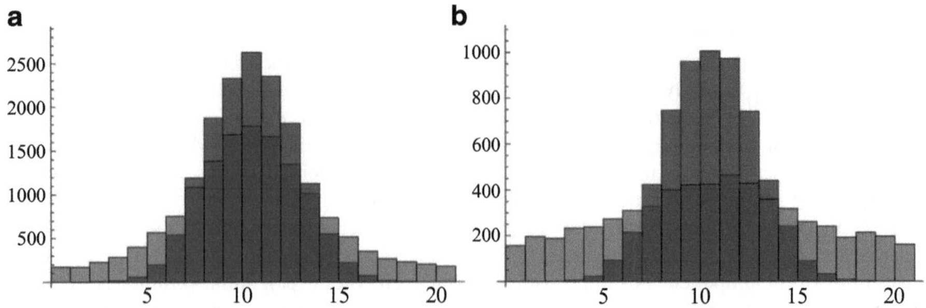

premisses and a false conclusion is only necessary, but not sufficient for deductive validity. In Part II, however, we shall, for convenience, refer to those arguments with a true conclusion or at least one false premiss as the "deductively valid" arguments.

Generally, this report's investigation follows the methodological KISS-principle[11]: We start by designing, as outlined above, very simple models of debate dynamics. Carrying out simulations based on these models will allow us to study which phenomena can be reproduced and thence, in principle, be explained by comparatively simple mechanisms. Subsequent simulation experiments will gradually replace the simplistic assumptions. The simple models initially studied may then serve as a foil with which we contrast later results.

[11]Cf. Hegselmann and Krause (2002, p. 9). "KISS" stands for "keep it simple, stupid".

Part I
Why Do We Agree? On the Consensus-Conduciveness of Controversial Argumentation

Chapter 3
Introduction to Part I

In Sect. 3.1, we outline the argumentative structure of Part I; we explain how its chapters, by exploring ever more sophisticated simulation setups, build on each other and follow a consistent line of reasoning. Based on this general orientation, we pinpoint, in Sect. 3.2, the different pieces of evidence which back the main results concerning consensus-conduciveness (cf. Sect. 1.4) and which are spread all over Part I. Hence, Sect. 3.2 shall serve as a bridge that connects the condensed results reported in the general introduction to the specific simulation studies and analyses carried out in the ensuing chapters.

3.1 Outline of Part I

We start, in Chap. 4, by studying the most simple implementation of the general simulation design. Accordingly, the simulation experiments presented in this chapter rely on highly simplified modeling assumptions. That is, there is no background knowledge, new arguments are constructed randomly, and proponents adopt the coherent position which is closest to their previous one. The simulation experiments confirm a basic hypothesis, which says that the overall consensus reached in a debate at some step t depends on two factors: (1) the initial agreement between proponents, and (2) the inferential density of the dialectical structure at step t. More specifically, the simulations reveal that, in general, proponents approach each other in a controversy (i.e., as the inferential density rises) only slowly—no matter whether we consider mean proponent agreement or the number of proponent positions which fully agree in a debate. Initial agreement influences the mutual rapprochement of proponents, too: proponents who have agreed broadly prior to a debate tend to agree throughout a debate, as well. However, initial agreement can also be destroyed during a controversy, as the simulations show. A random walk effect explains such alienation. In addition, we find that the doxastic dynamic of a controversy depends significantly on how newly introduced arguments shape the space of coherent positions (SCP). Debates with a highly fragmented SCP display, on average, a

G. Betz, *Debate Dynamics: How Controversy Improves Our Beliefs*,
Synthese Library 357, DOI 10.1007/978-94-007-4599-5_3,

different agreement evolution than debates with a compact SCP. We introduce the metaphors of the fishing net and, respectively, of the flooded village to explain the specific impact of the SCP's dynamic geometry. Besides giving rise to these detailed analyses, the simple simulation experiments highlight, quite generally, that engaging in a rational argumentation—where arguments are discovered randomly and there is no common background knowledge—in no way guarantees that proponents will approach each other, as long as the inferential density is not unrealistically high. This negative finding motivates further, more sophisticated simulation experiments, which are studied in the ensuing chapters: Does background knowledge on the one hand or the application of certain, purposeful (rather than random) argumentation strategies on the other hand allow proponents to reach agreement any faster than in the simple, basic setting? The investigations into these, arguably, more complex debate dynamics may expand on the very distinctions and explanations (e.g., the degree of fragmentation of the SCP) that helped to understand the dynamics of random debates.

In Chap. 5, we introduce background knowledge into the simulations by explicitly fixing the truth values assigned to some of the sentences, and investigate whether this fosters the rapprochement of proponent positions. Proponents are still assumed to adopt the closest coherent position, and arguments are discovered randomly. We find that a fixed body of background knowledge does not merely increase the agreement of the proponents by a constant amount. Rather, fixed background knowledge accelerates the rapprochement substantially, which can be explained by a gradual expansion of the so-called effective background knowledge in the course of a debate. Furthermore, the shape of the space of coherent positions clearly affects the position dynamics, albeit in an unexpected way. Unlike in the debates studied in Chap. 4, fragmentation turns out to be consensus-conducive, as, with background knowledge, proponents in fragmented debates reach mutual agreement more rapidly than those in compact debates. Yet, these observations can be understood by applying the explanatory framework of the flooded-village and the fishing-net metaphor.

Beginning with Chap. 6, we start to investigate whether the agreement evolution in a controversy depends on the way arguments are constructed and introduced. In other words, we replace the random argument construction, previous simulations relied on, by more elaborate argumentation rules. The analysis in Chap. 6 is restricted to dualistic debates, that is, debates between two agents. We consider four argumentation strategies the agents can pursue when introducing a new argument into the debate: "*fortify*," "*convert*," "*attack*," and "*undercut*." A proponent *fortifies* her own position, if she puts forward an argument containing but premisses and a conclusion she believes to be true. In the other three cases, the proponent considers, besides her own position, the stance of her opponent. She tries to *convert* the opponent if her argument (1) rests on premisses the opponent agrees with and (2) backs a conclusion adopted by the proponent. Arguing, based on premisses the proponent adheres to, against the position of the opponent, is referred to as *attacking* the opponent's position. If the new argument, finally, takes off from the opponent's position in order to demonstrate that a further conviction of the opponent, which does not figure as a premiss in this argument, is false, the opponent's position is *undercut*. *Fortify* and *attack* represent rather self-centered argumentation strategies, while *convert* and *undercut* are opponent-sensitive. Chapter 6 reports and discusses the agreement evolutions in dualistic debates where the proponents advance arguments in line with one of these argumentation strategies, scrutinizing the consensus-conduciveness of these rules. As a general result, we find that agreement evolution depends critically on the argumentation strategies employed. Opponent-sensitive strategies are significantly more consensus-conducive than self-centered ones. The specific performance of the rules in terms of mutual rapprochement can be explained with a view to how arguments, which are advanced in line with one of these rules, shape the space of coherent positions.

Having carried out a comparative analysis of different argumentation strategies by simulating dualistic (two-proponent) debates, we move, in Chap. 7, to many-proponent debates. Whereas in two-proponent debates with (basic) purposeful argumentation, at least every second argument is put forward with a view to a given proponent's position, this is only true for every sixth argument in a debate with six proponents (and, generally, every *m*th argument in a m-proponent debate). Thus, the greater the number of proponents, the more a controversy, where proponents follow one of the four basic rules, resembles a random debate. This is why the results of dualistic debate simulations cannot be directly scaled up to debates with more than two proponents. In particular, the high consensus-conduciveness of some strategies in dualistic debates doesn't warrant that these strategies are equally consensus-conducive in debates with six proponents, for example, those studied in Chaps. 4 and 5. The debates we simulate in Chap. 7 therefore contain six proponents who implement adjusted argumentation strategies which are derived from the basic rules introduced above. More specifically, we concentrate on two modifications of the *convert* and the *undercut* strategy, namely, *multiple convert* and *multiple undercut*. Both revised strategies are shown to be significantly more consensus-conducive than a *random argumentation* in comparable debates, while *multiple convert* clearly outperforms *multiple undercut*. The superior effectiveness in generating consensus can be explained with a view to the evolution of the SCP in such debates. First of all, *multiple convert* and *multiple undercut* lead to a substantially greater fragmentation of the SCP. And secondly, *multiple convert* succeeds in pushing together proponent positions on one and the same component of the SCP.

In the debates simulated in Chaps. 4–7, all sentences are on a par with each other. Proponents don't consider some of the debates' sentences as more and others as less important. If they can, for example, reestablish coherency by changing exactly one truth-value assignment, they are indifferent as to which belief they give up. But this, it seems, doesn't hold in real controversies, where proponents frequently possess some convictions which they are very reluctant to give up, as well as other beliefs they are much more willing to modify. In Chap. 8, we are going to include the proponents' varying loyalty to different beliefs in our simulations and adopt, in doing so, a Lakatosian perspective. More specifically, we assume that there is a subset of the sentence pool which contains the debate's core sentences. While the proponents still maintain complete positions that are defined on all sentences, their partial positions on the core sentences, we assume, make up the heart of their belief system, which they are particularly unwilling to modify. Technically, this translates into a more sophisticated update mechanism employed in the simulations. Our focus in Chap. 8 is not on the evolution of the proponents' complete positions, but turns to the proponents' core positions and, in particular, the agreement vis-à-vis the debate's core sentences. We find that, with *random argumentation*, the introduction of core beliefs, which proponents revise only reluctantly, causes the rapprochement in regard to these core sentences to slow down dramatically. As the proponent core positions represent partial positions defined on a subset of the sentence pool, we will be able to examine, moreover, how the robustness of positions, that is, their degree of justification, influences the debate dynamics. Robustness, it turns out, indicates proximity to a debate's final consensus. We propose to explain this by the comparatively strong immunity against future falsifications of core positions that possess a high degree of justification.

Chapter 8 distinguishes between core beliefs, which proponents give up only reluctantly, and auxiliary beliefs, beyond the debate's core, which proponents are much more willing to alter. It studies the effect of this Lakatosian distinction while retaining the simple *random argumentation* mechanism. Clearly, core beliefs can also be taken into account when putting forward new arguments. In Chap. 9, we devise two argumentation mechanisms which are sensitive to the difference between core and auxiliary beliefs. The first one is derived from the most effective argumentation strategy studied so far. The *multiple core convert* strategy tries to convert as many opponents as possible while explicitly targeting their core convictions. The design of the second argumentation mechanism, which we consider in Chap. 9, is motivated by the finding that a core's robustness exerts a significant influence upon the future evolution of the proponent's position. This suggests to maximize, as an argumentation rule, the robustness of one's core position. Both argumentation rules accelerate, as compared to random debates, core rapprochement. The consensus-conduciveness of *robust argumentation*, we argue, stems from its ability to shape the SCP in an appropriate, agreement-increasing way. Yet, employing these rules has opposite effects with a view to robustness: While robustness becomes more accurate an indicator of consensus proximity with *robust argumentation*, the *multiple core convert* strategy decreases the accuracy of this indicator.

3.2 Main Results and Their Justification

In the following, we reproduce, in slightly abbreviated form, the main results regarding consensus-conduciveness from Sect. 1.4 and point out which specific simulation experiments support them.

C1 (General Results) Controversial argumentation is, all things considered, consensus-conducive. Although the concrete agreement evolution in an individual debate seems to depend, mainly, on random factors, we may nonetheless discern substantial statistical differences between different argumentative practices.

We observe upward-sloping ensemble-wide mean agreement evolutions in (almost) every ensemble studied in Part I (cf. Figs. 4.1, 5.1, 6.1, 7.1, 8.2 and 9.2). Likewise, we find that, in nearly every simulation, the average number of fully agreeing proponent positions increases throughout the debates (or that, equivalently, the number of nonidentical proponent positions decreases, cf. Figs. 4.3, 5.4, 6.3, 7.2, 8.3 and 9.3). The variation of agreement evolutions within one and the same ensemble, that is, the fluctuation that is due to random factors, is illustrated and discussed in Sect. 4.2 (in particular Fig. 4.1a, see also Figs. 4.4 and 5.2).

C1.1 (Long Run) A controversial argumentation compels proponent positions to converge, eventually. Different argumentative practices vary substantially with respect to the pace of this convergence.

In the long run, that is, at a density of $D = 1$, only one single position remains dialectically coherent (cf. Sect. 2.5). Hence, proponents inevitably agree at this point of a debate. The distinct influence of initial and boundary conditions, or of argumentative practices, on the mean consensus evolution pertains to earlier phases of a debate, as the following results detail.

C1.2 (Alienation) Controversial argumentation may very well, in particular during the initial phase of a debate, lead to the alienation of proponent positions and undo coincidental agreement. This effect, too, depends strongly on the argumentative strategies employed by the proponents. It is, in line with (C1.1), inevitably reversed in the long run.

With *random argumentation*, high initial agreement evaporates in the first phase of a controversy (cf. Figs. 4.2 and 5.3). This can be explained by a random walk effect (see specifically Sects. 4.3 and 4.5, but also Sect. 6.3 and 7.3). Depending on the argumentative strategy employed by the proponents, the alienation effect can be softened or strengthened (see C3.1–C3.3 below).

C1.3 (Global Agreement Versus Partial Consensus) There exists a trade-off between (a) increasing the overall mean agreement between *all* proponents in a debate and (b) prompting at least some proponents to agree fully. Debate evolutions which foster partial consensus (i.e., full agreement between some proponents) tend to slow down the global rapprochement of proponent positions.

The trade-off is explicitly noticed in random debates (cf. Sects. 4.3–4.5). But it holds in the *multiple undercut* ensemble, too (cf. Fig. 7.3). We explain the trade-off with regard to the geometry of the space of coherent positions and, more precisely, by applying the fishing-net and the flooded-village metaphor, which illustrate distinct types of position dynamics.

C1.4 (The Space of Coherent Positions) The characteristic consensus dynamics of argumentative practices can be explained in terms of how the corresponding argumentation shapes the space of coherent positions. In particular, the degree of fragmentation of the space of coherent positions turns out to be of pivotal importance for the belief dynamics.

We distinguish, when plotting detailed simulation results, highly compact and very fragmented debates (e.g., Figs. 4.6, 5.3 and 7.3). Moreover, the concept of the space of coherent positions turns out to be a powerful tool for understanding the position dynamics (e.g., Fig. 4.5). Thus, we can explain, with a view to the evolving shape of the SCP, the trade-off reported above (C1.3), the effects of introducing background knowledge (see Sect. 5.3), the consensus-conduciveness of the basic argumentation strategies in dualistic debates (see Sect. 6.3, specifically Figs. 6.5–6.7), the outstanding consensus-conduciveness of *multiple convert* (cf. Fig. 7.5), and the role of robustness, that is, degree of justification, in the argumentative dynamics (see Sect. 8.3, specifically Fig. 8.6, and Sect. 9.3, Figs. 9.9 and 9.10).

C2 (Background Knowledge) The introduction of background knowledge into a debate fosters, very much as one would expect, the mutual agreement between proponents.

This finding is backed up by Sect. 5.2.

C2.1 (Multiplier Effect) The introduction of constant background knowledge accelerates the rapprochement of proponent positions.

Constant background knowledge does not only raise mean agreement by a fixed amount throughout a debate but speeds up the agreement increase (see, again, Sect. 5.2). This is because, as the debate unfolds, ever more sentences can be derived from the constant body of background beliefs. These sentences become, consequently, part of the so-called effective background knowledge themselves and may, in turn, serve as a basis for the derivation of further statements. This *multiplier effect* drives the discernible speedup of mutual rapprochement (cf. Sect. 5.3, in particular Fig. 5.6).

C2.2 (Favorable Fragmentation) With a sufficiently broad body of background knowledge, the fragmentation of the space of coherent positions, which tends to obstruct mean agreement increase without background knowledge (C1.4), favors both the generation of partial consensus and the global increase of mean agreement, thus resolving the trade-off reported above (C1.3).

We observe this particular effect in Sect. 5.2 (see Figs. 5.3 and 5.4).

C3 (Argumentation Strategies) The consensus-conduciveness of specific argumentative practices varies widely. The most noteworthy differences pertain to self-centered argumentation rules on the one side and opponent-sensitive ones on the other side.

Specific, purposeful argumentation strategies are studied, and compared to *random argumentation*, in Chaps. 6, 7, and 9.

C3.1 (Self-Centered Argumentation) Self-centered argumentation strategies, that is, argumentation rules (such as *fortify* and *attack*) which stipulate that a proponent advances but arguments with premisses she accepts as true, are totally ineffective in generating agreement. Strategies which are in addition aggressive, recommending direct attacks against opponent positions (e.g., the *attack* rule), consistently destroy agreement in all phases of a debate, and drive proponent positions systematically apart.

The poor consensus-conduciveness of the *fortify* and the *attack* rule is documented in Figs. 6.1 and 6.3. Moreover, we find, in Sect. 6.2 (Fig. 6.2), that the *attack* strategy pushes proponents away from each other irrespective of their initial agreement.

C3.2 (Opponent-Sensitive Argumentation) Opponent-sensitive argumentative practices are highly consensus-conducive. So, using only statements which an opponent considers true as premisses (of the arguments one introduces to back up one's position), represents the most effective way for generating agreement.

Figures 6.1 and 6.3 testify to the superior consensus-conduciveness of the *convert* and *undercut* rule. The result is, generally, corroborated by the study of *multiple convert* and *multiple undercut* (see Sect. 7.2, specifically Figs. 7.1 and 7.2), as well as by the investigation into *multiple core convert* (cf. Sect. 9.2, Figs. 9.2 and 9.3).

C3.3 (Aggressiveness and Disagreement) Aggressive opponent-sensitive strategies, that is, extremely critical strategies such as the *undercut* rule, are, in general, less consensus-conducive than their nonaggressive counterparts (*convert*). The less aggressive *convert* rule, moreover, allows for an apparently highly beneficial strategy: Before directly refuting an opponent position, potential back doors (adjacent fall-back positions) which are available to the opponent and which are farther removed from the proponent than the opponent's current position are closed (rendered incoherent). When the opponent position is, afterwards, directly refuted, the opponent is compelled to relocate towards the proponent.

Sections 6.2 and 7.2 reveal the superior performance of the less aggressive *convert* strategy (in its different versions). The agreement engineering process, outlined above, is identified and discussed in Sects. 6.3 and 7.3 (see, in particular, Fig. 7.5).

C3.4 (Friends and Fundamentalists) The effectiveness of an argumentation strategy in generating consensus depends on whether the initial agreement with one's opponent is very high ("friend") or very low ("fundamentalist"). Thus, a sharply critical, aggressive opponent-sensitive rule is advisable when arguing with a fundamentalist. Frequent falsifications due to "internal critique" represent in fact the most appropriate means for overcoming extreme dissent. Massive criticism impedes, however, finding consensus when arguing with a friend. Minor disagreement, instead of being effectively resolved, is typically deepened by aggressive argumentation. Here, the less critical *convert* strategy is much more consensus-conducive than the *undercut* strategy.

The interplay between argumentation strategies on the one hand and initial conditions on the other hand is depicted in Fig. 6.2 and discussed in Sect. 6.3 (but compare also Fig. 7.1).

C4 (Consensus Bias) Different argumentative practices do not only vary with respect to their consensus-conduciveness. The argumentation strategies employed by the proponents affect, in addition, the distance between the proponents' initial positions and the debate's final consensus.

Chapters 6, 8, and 9 explicitly observe and discuss the distance between proponent positions and the corresponding debate's final consensus.

C4.1 (Resilient Argumentation) The final consensus reached in a debate tends to be closer to the initial positions held by proponents who employ an opponent-sensitive argumentation strategy (i.e., the *convert* or *undercut* rule) than to the initial positions maintained by proponents who argue in a self-centered way (implementing the *fortify* or *attack* rule).

The versatility of proponent positions is a function of their associated argumentation rule, as shown in Fig. 6.4. We analyze, in Sect. 6.3, why proponents who employ the *convert* or *undercut* rule possess the most resilient, i.e. least versatile, positions.

C4.2 (Robust Core Positions) Proponent core positions with a high degree of justification at an early phase of the debate tend to be closer to the final consensus than core positions which exhibit a low degree of justification. This is because the higher the degree of justification, the more robust the corresponding core position and the more flexibly can a proponent adapt her complete position to critical arguments without modifying her core beliefs.

In random debates, proponent core positions which display a low robustness at an early stage of a debate are typically modified, in the ensuing debate, to a broader extent than positions which are relatively robust (cf. Fig. 8.4a). As we discuss in Sect. 8.3, degree of justification is, accordingly, an indicator of proximity to a debate's final consensus (see, in particular, Fig. 8.5).

C4.3 (Sensitivity of Indicator) Proponents who introduce arguments so as to maximize the robustness of their core position don't reach a consensus any faster than proponents who apply opponent-sensitive strategies. Yet, in debates where proponents maximize their robustness through argumentation, the accuracy of the degree of justification as an indicator of a position's agreement with the final consensus increases dramatically. In contrast, the correlation between robustness and agreement with the final consensus almost vanishes entirely if proponents pursue very aggressive and critical strategies.

The sensitivity of robustness as an indicator for durability (i.e., proximity to the final consensus) is studied in Chap. 9 (see Figs. 9.4 and, specifically, Fig. 9.8).

Chapter 4
The Consensual Dynamics of Simple Random Debates

The simulation experiments presented in this chapter rely on highly simplified modeling assumptions. That is, there is no background knowledge, new arguments are constructed randomly, and proponents adopt the coherent position which is closest to their previous one. These simulations will serve to scrutinize a simple hypothesis, which says that the overall consensus reached in a debate at some step t depends on two factors: (1) the initial agreement between proponents and (2) the inferential density of the dialectical structure at step t. We will try to gauge, roughly, which combinations of inferential density and initial agreement typically generate a full consensus. A more detailed analysis of the simulation uncovers that the dynamic geometry of the space of coherent positions exerts a pivotal influence on consensus evolution in a debate.

4.1 Setup

The specific set up of the simulations presented in this chapter is fairly simple. The pool S from which premisses and conclusions of arguments are drawn comprises $2 \cdot 20$ sentences ($n = 20$). The evolution of six different proponent positions, $\mathscr{P}^1, \dots, \mathscr{P}^6$, each being a complete position declared on the pool of sentences, is simulated ($m = 6$). The argumentation, discovery, and update mechanisms are as follows:

Argumentation Mechanism: At each time step, a new, random argument is added to τ. The new argument is constructed as follows: Two premisses and one conclusion are drawn randomly (and independently) from S; they make up a "candidate argument." If the sentences contained in the candidate argument are contradictory, if the conclusion is identical with a premiss, or if adding the candidate argument to the dialectical structure (adjusting the dialectical relations accordingly) has the effect that no complete position is coherent on the extended τ at all, the process is repeated and a new candidate is drawn randomly.

G. Betz, *Debate Dynamics: How Controversy Improves Our Beliefs*,
Synthese Library 357, DOI 10.1007/978-94-007-4599-5_4,

Otherwise, the candidate argument is added to τ_t, thus giving rise to τ_{t+1}. We shall refer to this mechanism as *random argumentation.*

Discovery Mechanism: The background knowledge is and remains empty.

Update Mechanism: Once τ_{t+1} is specified, it is checked (for every $i = 1 \dots 6$) whether $\mathscr{P}_t^i$ is coherent on τ_{t+1}. If it is, the position i remains unchanged ($\mathscr{P}_{t+1}^i = \mathscr{P}_t^i$). If it isn't, $\mathscr{P}_{t+1}^i$ is set to the closest coherent position to $\mathscr{P}_t^i$; that is, $\mathscr{P}_{t+1}^i$ is that position $\mathscr{P} \in \Gamma_{\tau_{t+1}}$ with minimal $\Delta(\mathscr{P}, \mathscr{P}_t^i)$. In case there are several closest τ_{t+1}-coherent positions, one of those is chosen randomly. Let us call this mechanism *closest coherent.*

The simulation terminates if the density of the dialectical structure equals 1, $D(\tau_t) = 1$ (note that the inferential density cannot exceed 1 because of the specific argumentation mechanism).

This kind of simulation of a single debate's evolution is carried out 1,000 times, giving rise to an ensemble of individual debate simulations.[1]

4.2 Results

Figure 4.1a depicts, for each debate in the ensemble, the evolution of the debate-wide mean normalized agreement between the proponents. Figure 4.1b aggregates the information contained in (a), showing how the normalized agreement—this time not only averaged over all position pairs in a debate, but additionally over all debates in the ensemble—evolves as a function of inferential density. At low inferential densities, mean normalized agreement is close to 0.5, which is the expected value given a purely random and unbiased distribution of proponent positions. Mean agreement rises slowly to roughly 0.55 as inferential density approaches a $D = 0.5$. Densities above this value cannot be realistically attained without a substantial body of implicit background knowledge, as detailed in the previous chapter. However, in the simulated random debates, significant rapprochement only kicks in as soon as inferential density increases well beyond 0.5. On average, full agreement isn't reached unless the inferential density equals one, that is, unless there is but one coherent position left.

The shaded area in Fig. 4.1b visualizes the ensemble-wide average ratio of τ-analytic sentences as a function of inferential density. In a given debate, the normalized agreement between any two positions is at least as great as the ratio of τ-analytic sentences (see Sect. 2.3). The thick curve lies, consequently, well above the gray area. However, it is noteworthy that agreement starts to increase significantly only once some sentences have become τ-analytic. The shape of the mean normalized agreement curve, moreover, seems to follow the shape of the evolution of the τ-analytic ratio.

[1]The software which is used to carry out the simulations of this study is documented at http://sourceforge.net/projects/ardys/

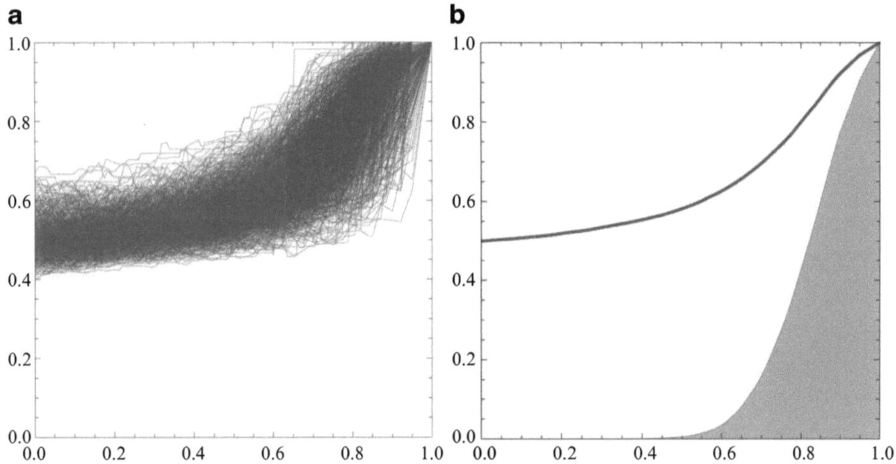

Fig. 4.1 Normalized agreement evolutions as a function of inferential density. In plot (**a**), every curve displays a debate-specific mean normalized agreement. In order to calculate debate-specific mean normalized agreements, normalized agreement $(1-\triangle)$ is averaged over all pairs of positions in that very debate. Plot (**b**) displays the evolution of the ensemble-wide mean normalized agreement (*thick line*) and the ratio of tau-analytic sentences (*shaded area*). Both variables are plotted against inferential density and averaged over all debates in the ensemble. The *thick line* thus depicts the mean of all the curves in plot (**a**)

Figure 4.1 plots mean agreement values, aggregated over individual debates, and, respectively, the entire ensemble. Two concrete proponent positions in some debate might therefore approach each other much faster or slower than indicated by the average curves of Fig. 4.1. In order to investigate whether positions with a high initial agreement converge faster and collapse sooner than two average positions, we classify, in Fig. 4.2, all position pairs that occur in a debate according to their initial agreement. More specifically, we distinguish three classes: pairs of proponent positions whose initial normalized agreement lies between 0 and 0.2 (great dissent), between 0.4 and 0.6 (medium agreement), and between 0.8 and 1 (high agreement). Like Fig. 4.1b, Fig. 4.2 plots the ensemble-wide mean normalized agreement, while limiting the aggregation to the three classes just specified. The top curve displays the mean normalized agreement for position pairs with high initial agreement; the bottom curve represents the agreement evolution of those position pairs with extreme initial disagreement. Let's have a closer look at each of the three curves. The intermediate curve closely resembles the ensemble-wide mean plotted in Fig. 4.1b. Position pairs whose initial agreement lies between 0.4 and 0.6 behave, on average, pretty much like the mean of all position pairs in the ensemble. This is apparently neither the case for those position pairs with high initial dissent, nor for those with high initial agreement. Concerning the positions that lie, initially, far apart, agreement begins to rise significantly as soon as the inferential density

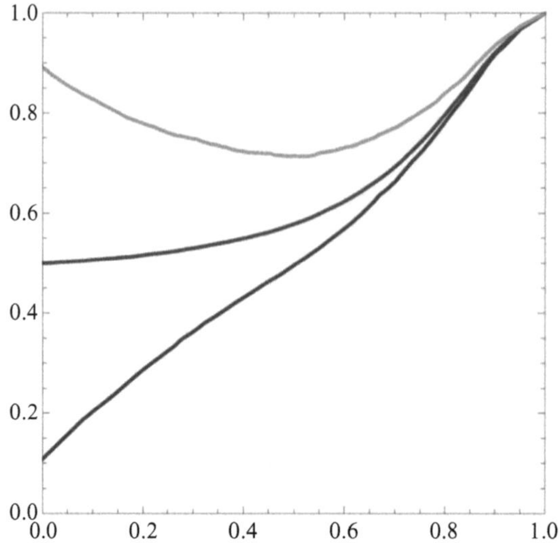

Fig. 4.2 Ensemble-wide mean agreement as a function of inferential density for different initial agreement conditions. The mean agreement represented by the *bottom curve* is calculated by taking only pairs of positions into account whose initial agreement lies between 0 and 0.2. Likewise, the intermediate curve averages over position pairs with medium initial agreement ($0.4 \leq (1 - \triangle) \leq 0.6$), and the *upper curve* depicts mean agreement for pairs of positions that initially agree on at least 80% of the sentences

is increased. Whereas the ensemble-wide mean agreement curve is virtually flat on the inferential density interval $[0; 0.5]$, the bottom curve in Fig. 4.2 displays a constant increase. When the inferential density has reached 0.5, proponents who initially agreed, on average, on 10% of the sentences in the debate have come to agree on more than 45%. Beyond an inferential density of 0.5, the rapprochement accelerates, and the bottom curve joins the intermediate one. This contrasts starkly with the dynamics of the position pairs which agree initially with regard to nearly all of the sentences, shown by the upper curve. They exhibit, probably, the most astonishing behavior. Taking off at a mean normalized agreement of roughly 0.9, the partial consensus evaporates as the inferential density increases. At a density of $D = 0.5$, the mean agreement has dropped by 0.15: Whereas proponents initially agreed on approximately 90% of the sentences that figure in the debate, they now agree on less than 75%. Only well beyond a density of 0.5, this trend is reversed. Increasing inferential density further eventually starts to foster agreement; but full consensus won't be reached before the density equals 1, either.

So far, we have described the debates' consensus dynamics merely in terms of the average normalized agreement between proponent positions. Counting the number of nonidentical positions opens an alternative perspective on studying consensus. The number of nonidentical (i.e., distinct) positions is, to a large extent, independent

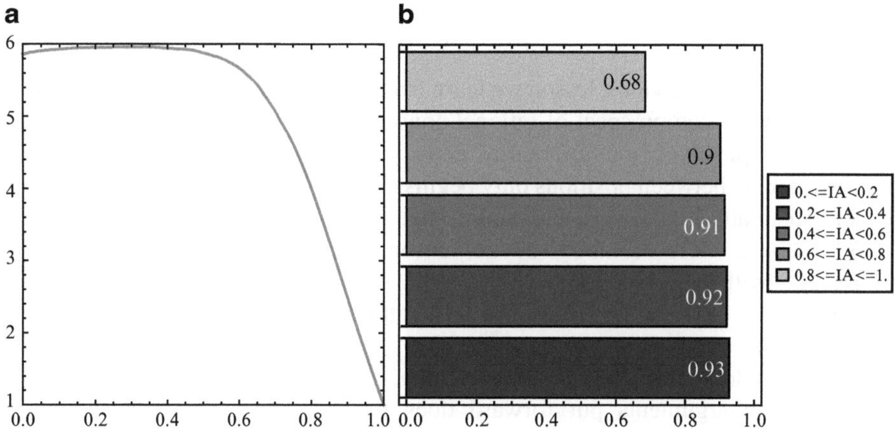

Fig. 4.3 (**a**): Ensemble-wide mean number of nonidentical positions in a debate as a function of inferential density. (**b**): Mean inferential densities at which position pairs with a given initial agreement come to agree

of the average agreement that pertains among the proponent positions: A high proportion of nonidentical positions is consistent with a small average distance and vice versa. Figure 4.3 gives an insight into the evolution of consensus, averaged over the ensemble, in terms of nonidentical positions. The left-hand plot shows how the number of nonidentical proponent positions declines as inferential density increases. Virtually all six positions in a debate remain distinct as long as the inferential density stays below $D = 0.5$.[2] Beyond that density, the number of nonidentical positions begins to fall and drops to 3 for a density of 0.8. In other words, if a debate has reached a density of 0.8, the number of nonidentical positions has halved (as compared to the initial state). The ratio of nonidentical positions finally continues its steep incline. Moreover, the inferential density at which two proponent positions typically come to agree depends on their initial agreement, as the right-hand panel (b) demonstrates. As expected, positions with high initial agreement collapse onto each other at significantly lower densities than positions which were initially far apart. Still, on average, two positions which disagree on less than 20% of the sentences won't come to full agreement unless the inferential density exceeds 0.68. Full agreement for positions with greater initial dissent requires an inferential density of roughly 0.9, or even higher.

[2]The initial incline is due to the sampling of the initial proponent positions (cf. footnote 10): With this sampling method, roughly 2.5% of the proponent positions initially agree. This contrasts with an unadjusted random sampling where the proportion of identical positions is negligible. So, in the first few steps of the random debate evolution, the purely coincidental agreements, which are due to the sampling method, are destroyed as proponents adjust, independently of each other, their positions to new arguments.

4.3 Discussion

The first, and most obvious, lesson we learn from the results reported above says that there is only limited prospect of rational consensus. Although agreement depends, on average, positively on inferential density and initial agreement, substantial rapprochement between positions only begins once inferential density has increased to unrealistically high levels. Presuming that arguments are introduced randomly into a debate, there seems to be no reason to expect proponent positions to converge—with one exception, namely, the significant mean rapprochement of fundamentally opposed positions. We will return to this result at the end of this chapter.

A second lesson is even more intriguing. Engaging in argumentation, being responsive to arguments put forward does not necessarily increase collective agreement. Argumentation, the simple simulations in this chapter demonstrate, may as well antagonize and alienate proponents. Instead of moving closer in the course of a debate, proponent positions can actually be pushed apart. The decrease of mutual agreement due to ongoing argumentation, which has been observed for position pairs with high initial agreement, might be explained by two different, compatible mechanisms, namely, (1) a random walk effect and (2) the fragmentation of the space of coherent positions.

Ad (1). If the initial agreement between two positions is high, this merely represents a coincidence in the sense of not being caused by any inferential constraints (which are, initially, absent). With the continuous introduction of arguments, the proponents' positions are, from time to time, revealed as incoherent and adjusted according to the update mechanism. As long as the inferential density is low, there are a several closest coherent positions which a proponent, whose position has been rendered incoherent, might adopt, and one of these will be chosen randomly. Because new arguments are introduced randomly; because moreover, two initially very close positions are not necessarily affected by the very same arguments; and because proponent positions are updated independently of each other, the ongoing argumentation leads to unrelated and—to a large extent—random, gradual modifications of the proponent's positions, which resemble random walks (on a hypercube, to be precise). During such a random walk, the purely coincidental high initial agreement vanishes. Unless the inferential constraints are sufficiently strong so as to channel the updating process, new (random) arguments merely trigger random modifications that undo contingent and above-the-average agreement—as well as, for that matter, coincidental extreme disagreement. This explains the initial fall of the upper curve as well as the immediate rise of the lower curve in Fig. 4.2.

Ad (2). The random walk effect is, however, not the only mechanism by which argumentation can decrease rather than increase agreement. We may discern a further mechanism in terms of how the introduction of additional arguments modifies the space of coherent positions. Proponent positions are located in the space of coherent positions. New arguments, modifying the space of coherent positions, may cause proponent positions to relocate in that space. The dynamics of proponent positions is a sporadic movement in a continuously changing space of Boolean vector. Decisively, new arguments may shape the space of coherent positions in various ways, inducing thereby very different movements of the proponent positions. In particular, a new argument might eliminate all coherent positions which previously linked two proponent positions so that additional arguments will henceforth cause the two positions to move in opposite directions rather than to approach each other. By fracturing the space of coherent positions, argumentation may create "opinion islands" which are divided by an ocean of incoherent positions and which may be populated by proponent positions. Introducing new arguments potentially shrinks these islands, thus increasing the distances even further. The update mechanism *closest coherent* ensures that proponent positions, when being forced to adjust to new inferential constraints, stay on such an isolated component of the space of coherent positions rather than jumping onto another component. Clearly, this mechanism does not only apply to positions that exhibit high initial agreement. It is thus not suited to explain why, as regards the ensemble-wide average, only positions with high initial agreement tend to withdraw from each other. If the kind of alienation due to fragmentation sketched above is at work in the random debates, it applies potentially to all position pairs irrespective of their initial distance. We will pursue this idea in the following.

The hypothesis that the effect of argumentation on mean agreement depends on the dynamic geometry of the space of coherent positions can be put more precisely as follows. The compactness (or the fragmentation) of the space of coherent positions as well as whether proponents adopt central (or radical, far-off) positions in that space crucially determines the overall evolution of agreement and consensus in that debate.

We already possess a quantitative measure for (1) the compactness of the space of coherent positions and (2) the centrality of a specific position within that space, namely, normalized closeness centrality (NCC). The higher an individual position's NCC, the more compact the space of coherent positions and the more central the position's location within that space. Averaging over all positions and all time-steps in a debate gives us a measure of the overall compactness of the space of coherent positions in the sense of our hypothesis.

So, how do we expect debates whose space of coherent positions remains compact throughout the argumentation to behave? If a debate possesses, on average, a high aggregated NCC, the space of coherent positions shrinks gradually without fracturing and breaking up into different unconnected components. The space of coherent positions remains a set of closely related positions which is continuously "cut back" from its outer boundaries, so to say. Such an evolution of the space of coherent positions has two notable effects. First of all, the proponent positions, being located in the constantly contracting space of coherent positions, will gradually, and in small steps, approach each other. Secondly, full agreement between proponent positions won't be reached unless the inferential density is very high. This is because, provided the space of coherent positions remains interconnected, the proponent positions will possess a wide room of maneuver when being updated. Consequently, full agreement will be coincidental and unstable unless the number of coherent positions has become very small indeed. In sum, a perfectly compact evolution of the space of coherent positions can be likened, for illustrative purposes, to pulling a fishing net. The volume enclosed by the net corresponds to the space of coherent positions, and the fishes caught within the net represent individual proponent positions. As the net is pulled, the volume embraced gradually decreases, and the fishes are forced to move closer and closer; they will however, occupy one and the same position only once the net is fully tightened, and the volume is reduced to a minimum.

Let us, next, consider the opposite case, that is, a debate evolution in the course of which the space of coherent positions thins and stretches, and eventually fractures, while proponents tend to adopt extreme positions, situated near the outer boundaries of the space of coherent positions. The average NCC of such a debate is small. What kind of position dynamic does such a debate exhibit? If the space of coherent positions is thinned, or even fractured, ongoing elimination of coherent positions might drive proponents positions apart, leading, initially, to further antagonization and, possibly, radicalization. The space of coherent positions will consist in several distant, internally connected clusters of coherent positions that may or may not be populated by proponent positions. Proponent positions will remain on these gradually contracting components until the only coherent position left of such a component will have become incoherent, too. This forces the corresponding positions to adjust dramatically by modifying many truth-value assignments so as to "jump" onto the nearest remaining cluster of coherent positions. We may thus expect the average agreement to decrease initially—or at least not to increase—and then, later, to increase suddenly in one or several steps. Each such step corresponds to a collapse of an isolated and populated component of the space of coherent positions. Depending on the specific shape of the fragmentation, the number of nonidentical positions might be reduced significantly while the average disagreement grows: If several proponent positions are located on one and the same isolated component, shrinking this component might cause these positions to collapse onto each other although the entire space of coherent positions is still comparatively large. A sudden flooding of a village provides an, admittedly violent, but nevertheless apt, analogy to illustrate such a position dynamic. The village's non-flooded area corresponds to the space of coherent positions, the village's inhabitants to the proponent positions. As the water mark rises (inferential density increases), the streets and gardens will be flooded so that the inhabitants will have to search shelter on the upper levels of their houses, and eventually on the roofs. The nonflooded area has been fragmented. On average, people might have been driven apart by this process; in the same time, however, some might have found shelter on the very same roof, thus occupying identical positions. As soon as the first roofs are flooded as well, the poor inhabitants will have to swim to the remaining roofs. This starts to bring them together, decreasing the average distance in steep steps. In the end, the only remaining dry place might be the castle's roof and tower, where, after a series of dramatic relocations, the whole village ends up—and is, eventually, rescued.

Before we try to substantiate our hypothetical sketch of those two different types of position dynamics by a more detailed analysis of the simulation results, we should note that, according to the picture outlined above, there might actually be a trade-off between increasing mean normalized agreement on the one side and decreasing the number of non-identical positions (i.e., fostering complete agreement) on the other side. Whereas a continuous contraction of the space of coherent positions is favorable with regard to increasing the average agreement between proponent positions, it is somehow detrimental to reaching full consensus between some proponents. Fracturing the space of coherent positions and creating isolated "opinion camps" might be more effective with a view to causing at least some proponents to agree more or less fully. In other words, group cohesion might be achieved at the expense of overall agreement (except that if, by rare chance, all proponent positions are located, initially, in the same coin of the space of coherent positions, cutting off that coin from the rest of the space of coherent positions and gradually shrinking that isolated component might represent the most effective way for increasing mean agreement *and* for fostering full agreement).

4.4 Results, Continued

In order to scrutinize the hypotheses developed in the previous section, we distinguish, in our ensemble, debate simulations that possess, on average, a relatively high and, respectively, low average NCC. In particular, we calculate each debate's aggregated NCC by averaging the individual positions' NCCs over (1) all proponent positions and (2) all time steps with a corresponding inferential density smaller than 0.5. The following, more detailed analysis of the simulation results focuses on the upper and lower 10th aggregated NCC quantile of the debates in our ensemble (i.e., 10% of the simulation runs with the highest, and 10% of the runs with the lowest average NCC). We will explore whether debates with extremely high aggregated NCC (compact evolution of space of coherent positions) and extremely low aggregated NCC (fragmentation of space of coherent positions) display the features hypothetically spelled out above.

Figure 4.4 plots the debate-specific agreement evolution of the 100 most fragmented (a), and the 100 most compact (b) debates in our ensemble. In spite of the considerable variance, it is clear from this figure that the two types of debates behave differently. Before we consider the average features of the two classes, however, we will have a closer look at two illustrative examples. Figure 4.5 provides a detailed picture of the debate-specific mean agreement evolution and the corresponding change of the space of coherent positions for two individual debates, a compact one, belonging to the upper 10th aggregated NCC quantile, and a fragmented one with an aggregated NCC in the lower 10th quantile. Consider the scattered debate evolution (blue curve, lower snapshots) first. Mean agreement, averaged over the six proponent positions, rises slowly and sporadically from an initial value of 0.55 to 0.65, while inferential density increases to 0.4. Then, however, agreement drops abruptly and stays essentially on the initial level for densities up to 0.65. In two steep steps (at inferential densities 0.65 and 0.85), full agreement is eventually reached.

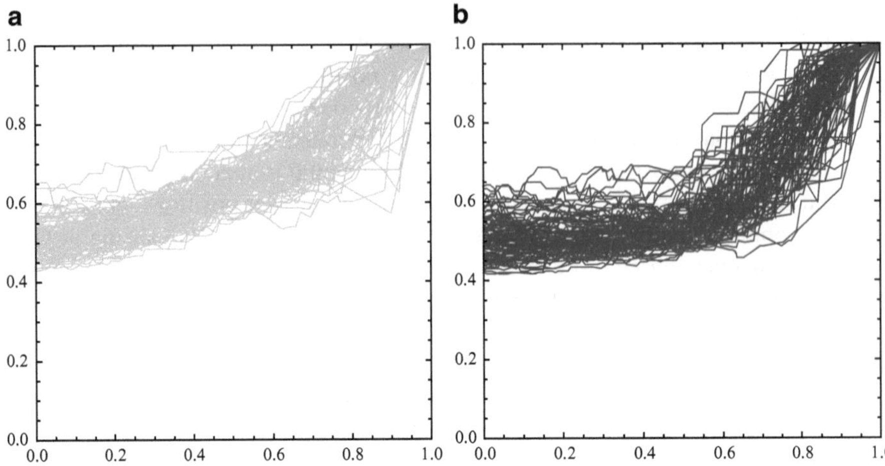

Fig. 4.4 Debate-specific mean agreement evolutions as functions of inferential density. (**a**): Compact debates, that is, debates with an aggregated NCC higher than the upper 10th quantile. (**b**): Dispersed debates, that is, debates with an aggregated NCC smaller than the lower 10th quantile

This uneven evolution of mean agreement corresponds to an uneven transformation of the space of coherent positions. At step 19, corresponding to an inferential density of 0.187, the coherent positions are well connected as shown by the 12-dimensional snapshot. The next snapshot, taken at step 29, displays, however, first signs of fragmentation. The space of coherent positions falls into two internally well-connected parts of roughly equal size, which are vertically separated by a sparsely connected area. The left-hand section is populated by two, and the right-hand section by four proponent positions. In the following, as shown by the snapshots at step 35 and 39, the connections between these two sections become ever sparser. Moreover, at step 39, an isolated section (consisting of two coherent positions) has been cut off from the main body. At step 45, a further chunk of the space of coherent positions has been detached, this one being populated with a proponent position. Two of the remaining proponent position are located at the far left of the main body, opposed to three positions at the far right. At step 46, the space of coherent position breaks up even further. It now consists of four isolated fragments—opinion islands—three of which are populated by proponent positions. Elimination of these fragments by further argumentation and contraction of the space of coherent positions will, eventually, bring the proponents together. Let us now turn to the compact debate evolution, portrayed by the green curve and the upper snapshots. In that debate, mean agreement takes off at an initial value slightly above 0.5 and rises in two phases, namely, at densities 0–0.3 and 0.4–1, with a relatively constant slope to full agreement. This steady increase is only interrupted by a short interval where mean agreement roughly stays constant. Taking into account the differences in initial agreement, the compact debate generates—compared to the fragmented one—fast

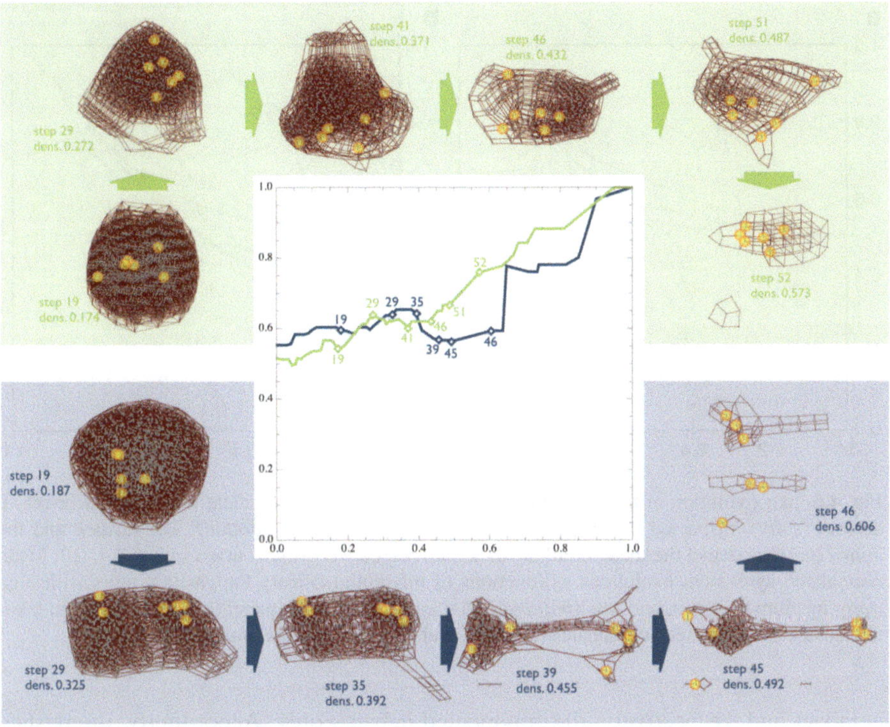

Fig. 4.5 Debate-specific mean agreement and space of coherent positions for two debates in the ensemble. Mean agreement is plotted against inferential density. The different graphs are 12-dimensional sections of the (20-dimensional) space of coherent positions. The *green curve* and the *upper snapshots* visualize the evolution of a compact debate; the *blue curve* and the *lower snapshots*, in contrast, depict a dispersed debate evolution. The time steps of the snapshots of the space of coherent positions are marked in the diagram. *Yellow circles* spotlight positions that are occupied by at least one proponent

and sustainable mean agreement. As illustrated by the corresponding snapshots, the space of coherent positions remains tightly packed and well connected for densities up to 0.5. Instead of being cut down from within, the space of coherent positions shrinks as if only its outer layers were gradually severed. Only at step 52, at a density well beyond 0.5, an isolated section emerges. Even at that point, however, the six proponent positions stay on the well-connected main body of the space of coherent positions. So, the general hypothesis about different kinds of debate evolutions, which we have articulated in the previous section, is nicely reflected and substantiated by these two illustrative cases. Next, we will investigate whether the overall, aggregated picture confirms that hypothesis, as well.

Figure 4.6 depicts how ensemble-wide mean agreement evolves as a function of inferential density. The solid lines represent the mean evolutions averaged over all debates and thus correspond to the plots in Figs. 4.1b and 4.2. The dotted and the dashed curves depict, in contrast, the mean agreement evolutions averaged over the

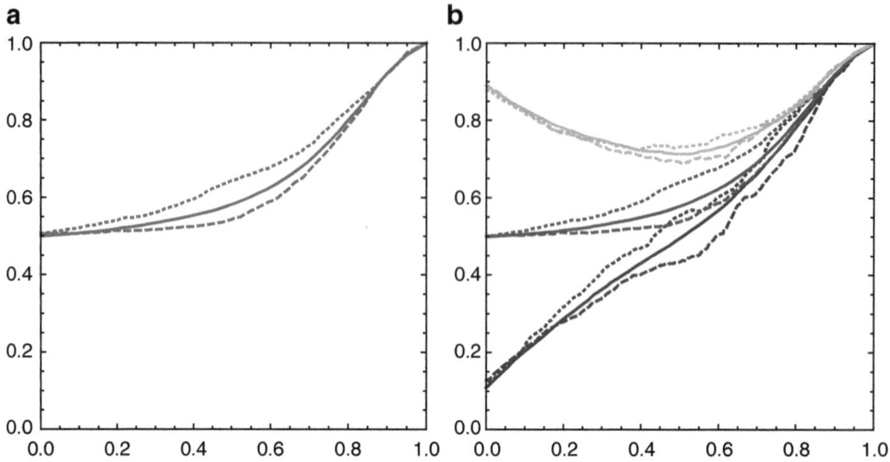

Fig. 4.6 (**a**): Evolution of the ensemble-wide mean normalized agreement taking into account all debates (*solid*), dispersed debates (*dashed*), and compact debates (*dotted*). The *dashed* and the *dotted line* thus depict the mean of all the dark and, respectively, light curves in Fig. 4.4. (**b**): Mean normalized agreement evolutions as functions of inferential density for position pairs with high (*top*), medium (*middle*), and low (*bottom*) initial agreement. As in panel (**a**), the calculation takes all debates (*solid*), dispersed debates (*dashed*), and compact debates (*dotted*) into account

compact and, respectively, the fragmented debates only. Accordingly, the dashed curve in Fig. 4.6a represents the average of all curves in Fig. 4.4a, whereas the dotted curve pictures the average of Fig. 4.4b. In compact debates, mean agreement (dotted curve in Fig. 4.6a) begins to rise notably at low densities and increases more steadily compared to the average debate (solid line). At a density of 0.5, compact debates exhibit a mean agreement of roughly 65%—almost 10 percentage points above the ensemble-wide average. In fragmented debates, however, agreement hardly increases for densities below 0.5 at all (dashed curve). The mean agreement at a density of 0.5 amounts to less than 55%. Beyond that density, though, the agreement evolution accelerates and catches up, in a comparatively steep rise, with ensemble-wide mean agreement (solid curve). This general picture—compact debates possessing higher mean agreement than fragmented ones, with the latter showing significantly slower rapprochement in the first half and a steeper rise of mean agreement in the second one—stays the same when we take different kinds of initial conditions into account. As can be seen in Fig. 4.6b, position pairs with medium initial agreement closely resemble the curves averaged over all position pairs, irrespective of their initial agreement (panel a). Position pairs with high initial disagreement approach each other quickly. In compact debates (bottom dotted curve), this rapprochement is even faster than in fragmented debates (bottom dashed curve). At a density of 0.5, mean agreement has increased by roughly 45 percentage points in compact as opposed to 30 percentage points in fragmented debates. The differences between compact and fragmented debates are least visible for position

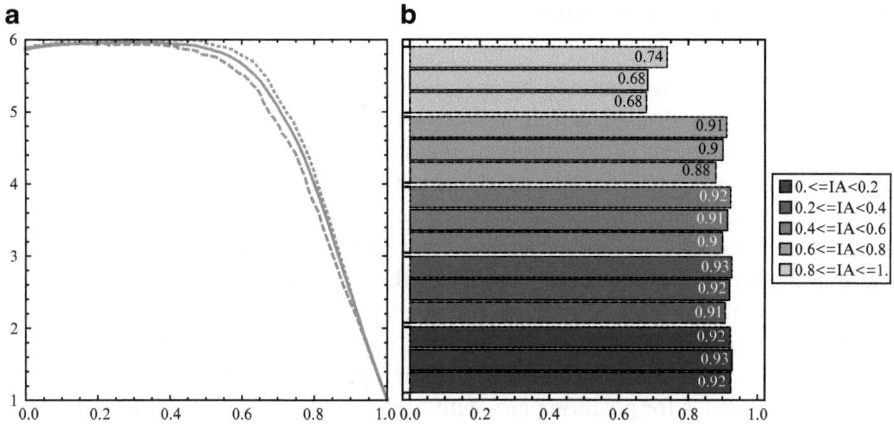

Fig. 4.7 (**a**) Ensemble-wide mean number of non-identical positions in a debate as a function of inferential density, averaged over all debates (*solid*), dispersed debates (*dashed*), and compact debates (*dotted*). (**b**) Mean inferential densities at which position pairs with a certain initial agreement (indicated by *shading*) come to agree, averaged over (1) all debates (*solid borders*, *middle*), (2) dispersed debates (*dashed borders*, *bottom*), and (3) compact debates (*dotted borders*, *top*)

pairs with high initial agreement (top curves). For low densities, the mean agreement evolutions lie very close to each other; only at a density of 0.4, mean agreement in compact debates (top dotted curve) starts to diverge, turning around and beginning a slow incline, whereas mean agreement in fragmented debates (top dashed curve) continues to fall until a density of 0.5 is reached. Further on, mean agreement evolutions quickly converge against each other.

Agreement evolution in terms of nonidentical positions is shown by Fig. 4.7, which corresponds to Fig. 4.3. As the left-hand panel shows, the number of nonidentical positions is nearly the same in compact (dotted line) and fragmented (dashed line) debates for densities below 0.4. Beyond this density, however, fragmented debates possess, on average, slightly less different proponent positions than compact ones. When there are, for instance, five different positions in an average fragmented debate (at a density of roughly 0.67), a typical compact debate possesses 5.5 nonidentical positions. In other words, two positions have collapsed in *every* fragmented debate, as opposed to *every second* compact debate. This difference gradually vanishes as full agreement is approached. Panel (b) of Fig. 4.7 displays the inferential densities at which position pairs with different initial agreement collapse, on average. Bars with dashed (dotted) borders depict the collapse densities in fragmented (compact) debates. As we had observed above, position pairs with high initial agreement collapse earlier than positions that lie, initially, further apart. Compactness or fragmentation of a debate's space of coherent positions appears to have only a marginal influence on collapse densities. In fragmented debates (dashed borders), position pairs seem to collapse slightly earlier than in compact debates. Merely for positions with high initial agreement (light shading), this difference is, however, clear and distinct.

4.5 Discussion, Continued

The aggregated picture outlined by the previous results nicely dovetails with our hypothetical distinction of different kinds of debate dynamics, characterized by compact and fragmented evolutions of the space of coherent positions. Compact debates give rise to a gradual and steady increase in mean agreement. In fragmented debates, however, proponent positions get caught in different, loosely connected segments of the space of coherent positions and only approach each other once these components are completely resolved, that is, at higher densities. Besides a confirmation of this general picture, the results presented in the previous section exhibit two peculiarities on which we shall comment next. The first peculiarity consists in the small difference between mean agreement evolution in compact and fragmented debates for position pairs with high initial agreement (cf. top curves in Fig. 4.6b); the second lies in the fact that, according to Fig. 4.7, fragmented debates generate agreement faster than compact debates.

Let us consider initially closely related position pairs, first. Why does a debate's degree of fragmentation hardly affect their agreement evolution? To understand this, we shall spell out what exactly we expect to see according to our hypothesis. Take a compact debate first. In such a debate, two proponent positions which agree initially (and purely coincidentally) to a high degree possess, thanks to the compactness of the space of coherent positions, a wide room of maneuver when being readjusted in the light of newly introduced arguments. The random walk mechanism will thence be able to fully unfold and drive the proponent positions apart before the inferential constraints are tight enough to bring the positions, eventually, together. Quite the opposite is true in a fragmented debate, or so it seems. There, the fragmentation of the space of coherent positions appears to limit the room of maneuver of closely related proponent positions early on. As the clustering tends to restrain the random walk effect, catching nearby positions on opinion islands, we'd expect proponent positions with high initial agreement to depart from each other to a lesser degree in fragmented than in compact debates. So, what is in need of an extra explanation is not the observation that the top dotted line lies only marginally above the top dashed line, but that it lies above it all, and not underneath. The explanation I may offer is this: The characteristic geometry of the space of coherent positions (compact or fragmented) emerges only as sufficiently many positions have been rendered incoherent, that is, once the inferential density is sufficiently high (compare the first snapshots in Fig. 4.5). Consequently, for low densities—say, up to 0.3—the space of coherent positions stays compact both in the debates with high- and in the debates with low aggregated NCC. As a result, proponent positions will have wide margins of maneuver when readjusting no matter whether they fall into compact or fragmented debates, and the random walk effect will fully kick in—driving down mean agreement in a similar way. But once the characteristic shape of the space of coherent positions emerges, the proponent positions, which initially were closely related, have already been driven far apart and are therefore as likely to end up on different fragments of the space of coherent positions as other position pairs. Gradual contraction of the space of coherent positions thus seems to be more effective in bringing the proponent positions together again than splitting up the space of coherent positions and shrinking the individual segments. That's why the top dotted line lies, for densities higher than 0.3, slightly above the dashed one.[3]

The second peculiarity, that is, the faster agreement in fragmented debates as depicted in Fig. 4.7, coheres well with our general hypothesis regarding the different debate dynamics. Far from contradicting the result that mean agreement increases more swiftly in compact debates, the fact that proponent positions in fragmented debates tend to collapse earlier than in compact debates can be explained by the specific dynamic geometry of the space of coherent positions. Thus, this result points to the opinion island effect hypothesized in Sect. 4.3. In a fragmented debate, proponent positions will be located on relatively small and fairly isolated segments of the space of coherent positions. As these fragments are contracted further, proponent positions lose any room of maneuver and are forced on identical positions even though the entire space of coherent positions still remains vast. Consequently, positions in fragmented debates collapse onto each other and reach

---

[3]This explanation suggests the following further test. If we partition all position pairs according to their agreement at a density of, say, 0.4 (rather than according to their initial agreement), the position pairs with high agreement should also possess a high likelihood of ending up on the same cluster, provided the debate is fragmented, and thence converge much faster than the corresponding position pairs on a compact debate. There are, in our ensemble, 78 position pairs in fragmented debates and 76 position pairs in compact debates which agree, at a density of 0.4, by more than 80% of the sentences. The evolution of the average normalized agreement of these two sets of position pairs is depicted in the following plot.

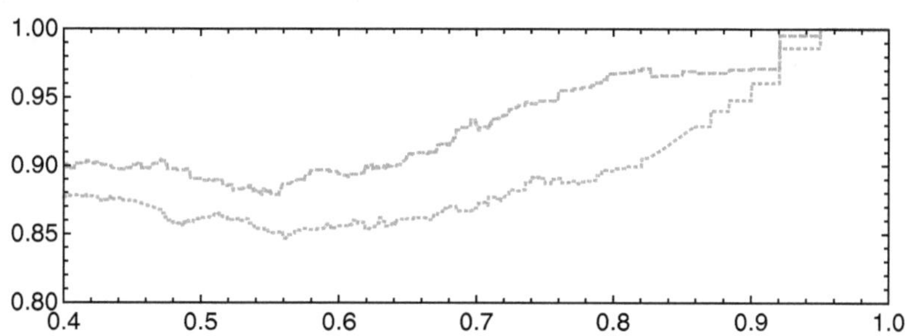

The dashed line represents the mean agreement evolution of the closely related position pairs in fragmented debates; the dotted line visualizes the corresponding evolution in compact debates. Both in fragmented and in compact debates, mean agreement drops as the density rises beyond 0.4 to densities between 0.5 and 0.6. In fragmented debates, this decline is turned, at a density of 0.6, into a rise which ends abruptly at a density of 0.8, having reached a mean agreement level well beyond 95%. In compact debates, in contrast, mean agreement slowly declines until a density of roughly 0.6 and, subsequently, starts a rather slow and gradual increase which catches up with mean agreement in fragmented debates—taking into account the initial difference at a density of 0.4—no earlier than at a density of 0.9. The observable, slightly more rapid mean agreement increase in fragmented debates can be attributed to the opinion island effect, which causes closely related position pairs, caught on one and the same isolated segment of the space of coherent positions, to approach each other as this segment is gradually contracted.

full agreement even at densities where their counterparts in compact debates have comparatively broad leeway and thus stay distinct. The results in Fig. 4.7, together with the fact that mean agreement increases more rapidly in compact debates, hence substantiate the alleged trade-off, referred to in Sect. 4.3, between increasing mean agreement among all proponent positions on the one hand and bringing about the full agreement between some proponent positions on the other hand. The former is more effectively achieved in compact debates, and the latter is more likely to occur in fragmented debates.

Given the discussion of the details and particularities of the simulation results presented hitherto, what are the general lessons to be learned from this chapter? First of all, the results of this chapter establish that argumentation does not necessarily generate consensus. Quite the opposite, argumentation may antagonize and alienate as well. Moreover, we have studied and understood two general mechanisms that are responsible for such an alienation: the random walk effect, and the pivotal role of the dynamic geometry of the space of coherent positions. Secondly, given the assumptions of the simulations presented in this chapter, there is actually only a remote chance for reaching consensus due to rational argumentation. It is an essential task of the following chapters to explore whether the *random argumentation* simulations can be modified so that rational consensus, that is, agreement which is due to the exchange of arguments, becomes viable. Generally, we may distinguish two levers in order to do so:

Introduce explicit background knowledge: By stipulating the (inter subjective) discovery of new facts, premisses or conclusions become shared background knowledge and provide a common foundation for the controversial argumentation.

Dialectically engineer agreement: One devises effective and non random argumentation strategies that further consensus and bring about significant agreement without necessarily increasing the inferential density beyond 0.5.

Chapter 5
The Consensual Dynamics of Random Debates with Explicit Background Knowledge

The simulations presented in the last chapter suggest that the overall prospects of reaching agreement due to argumentation are bleak as long as arguments are introduced randomly and as long as there are no commonly agreed upon background beliefs. In the following, we take background knowledge into account by explicitly fixing the truth values assigned to some of the sentences, and investigate whether this fosters the rapprochement of proponent positions.

5.1 Setup

Argumentation Mechanism: Arguments are introduced in accordance with *random argumentation* (cf. Sect. 4.1), which is supplemented by the additional verification that the new argument does not render the background knowledge incoherent.

Discovery Mechanism: The background knowledge $\mathscr{B}_t$ fixes the truth values of a specific proportion β (namely, 10%, 20%, and 40%) of the n sentence pairs in the sentence pool. It remains constant throughout the debate simulation.

Update Mechanism: Positions are updated according to a modified *closest coherent* mechanism (cf. Sect. 4.1), taking background knowledge into account. More specifically, once τ_{t+1} is determined, it is checked (for every $i = 1 \dots 6$) whether $\mathscr{P}_t^i$ is coherent on τ_{t+1} and extends $\mathscr{B}_{t+1}$. If it does, the position i remains unchanged ($\mathscr{P}_{t+1}^i = \mathscr{P}_t^i$). If it doesn't, $\mathscr{P}_{t+1}^i$ is set to the closest coherent position to $\mathscr{P}_t^i$ which extends $\mathscr{B}_{t+1}$; that is, $\mathscr{P}_{t+1}^i$ is that position $\mathscr{P} \in \Gamma_{\tau_{t+1}}(\mathscr{B}_{t+1})$ with minimal $\Delta(\mathscr{P}, \mathscr{P}_t^i)$. In case there are several closest τ_{t+1}-coherent positions, one of those is chosen randomly. Let us call this mechanism *closest coherent with background knowledge*.

G. Betz, *Debate Dynamics: How Controversy Improves Our Beliefs*, Synthese Library 357, DOI 10.1007/978-94-007-4599-5_5,

Initial proponent positions are assigned so as to be consistent with (i.e., so as to extend) the background knowledge. A debate contains six proponent positions. Three ensembles with $\beta = 0.1$, $\beta = 0.2$, and $\beta = 0.4$ are generated, each containing 1,000 individual debate simulations.

5.2 Results

Figure 5.1 plots the mean agreement evolutions corresponding to the three ensembles. Given that all positions agree at least with regard to the sentences which belong to the background knowledge, two randomly assigned positions differ, on average, with respect to half of the sentences which are not included in the background knowledge. This is the reason why ensemble-wide mean agreement evolutions take off at an initial value equalling $0.5 + \beta/2$. As the left-hand plots illustrate, a controversial argumentation becomes significantly more effective in terms of generating agreement once a shared background knowledge is established. At a density of 0.5, mean agreement has increased, relative to its initial value, by roughly 15 percentage points for $\beta = 0.1$. If the background knowledge comprises 20% or 40% of the sentences, argumentation raises the initial agreement even by more than 20 percentage points. For $\beta = 0.4$, proponents agree on average, at a density of 0.5, with respect to almost 95% of all the sentences. *Random argumentation* without background knowledge, in contrast, is merely able to raise the ensemble-wide mean agreement by significantly less than 10% at a density of 0.5 (cf. Fig. 4.1).

Besides raising, in general, the mean agreement evolution relative to initial agreement, debates with common background knowledge exhibit substantial rapprochement even at very low densities. Whereas, in random debates without background knowledge, mean agreement hardly increases for $D < 0.2$ at all, it rises immediately and almost linearly (in the density interval $[0; 0.5]$) for $\beta = 0.4$.

Ensemble-wide mean agreement evolutions of position pairs with a specific initial agreement, shown in the right-hand plots of Fig. 5.1, reflect the overall tendency: Broader background knowledge triggers more substantial and faster rapprochement. This implies, in particular, that position pairs with high initial agreement distance themselves from each other to a lesser degree as background knowledge widens. A background knowledge as large as 40% entirely prevents the alienation of proponent positions with coincidentally high initial agreement. Still, positions with low initial agreement approach each other notably faster at low densities than positions with high initial agreement.

The introduction of background knowledge has different effects on compact as opposed to fragmented debates. Figure 5.2 plots, as an overview, the debate-specific agreement evolutions of very compact and very fragmented debates in our three ensembles. As in the previous chapter, we consider debates in the upper and lower 10th quantile of aggregated NCC. The corresponding ensemble-wide mean agreement evolutions, both for all position pairs (left-hand panels) and for position pairs with specific initial agreement (right-hand panels), are displayed in Fig. 5.3.

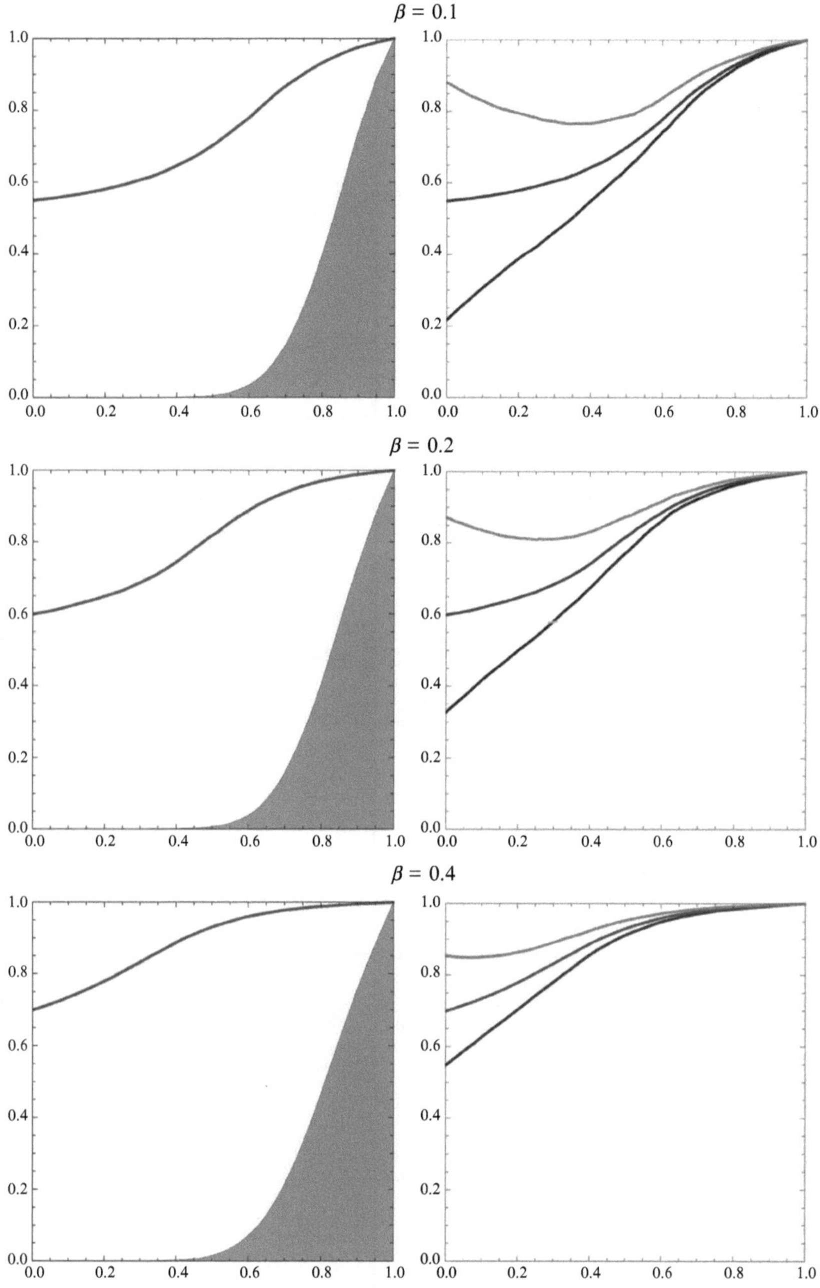

Fig. 5.1 *Left-hand panels*: Ensemble-wide mean agreement evolutions averaged over all position pairs, and proportion of tau-analytic sentences (*shaded area*). *Right-hand panels*: Ensemble-wide mean agreement evolutions of position pairs with different initial agreement; initial agreement intervals based on which the curves are calculated depend on β; they are, from *bottom* to *top*, $[\beta;\beta+0.2]$, $[0.4+\beta/2;0.6+\beta/2]$ and $[0.8;1]$

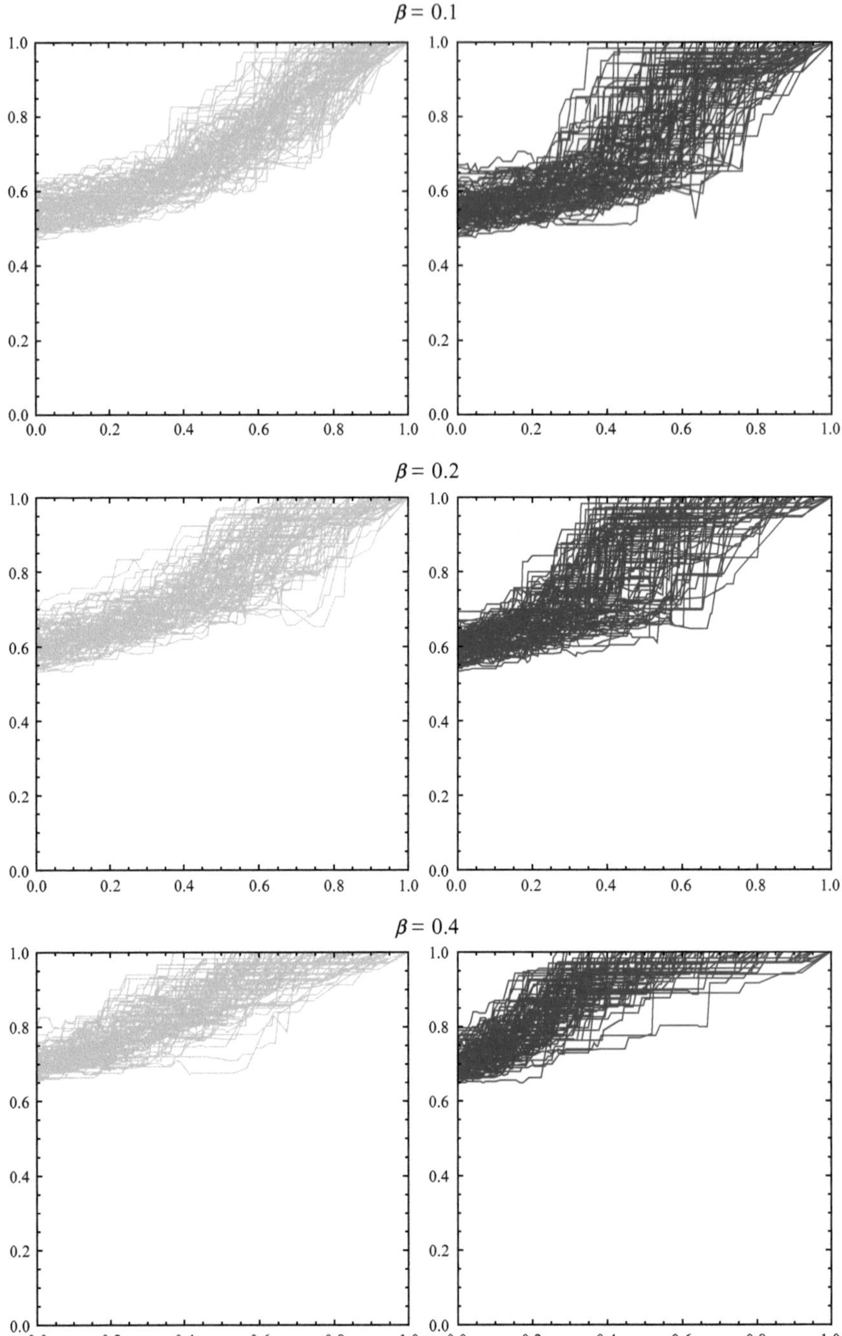

Fig. 5.2 Debate-wide mean agreement evolutions. *Left-hand panels*: compact debates with an aggregated NCC above the upper 10th quantile. *Right-hand panels*: fragmented debates with an aggregated NCC less than the lower 10th quantile

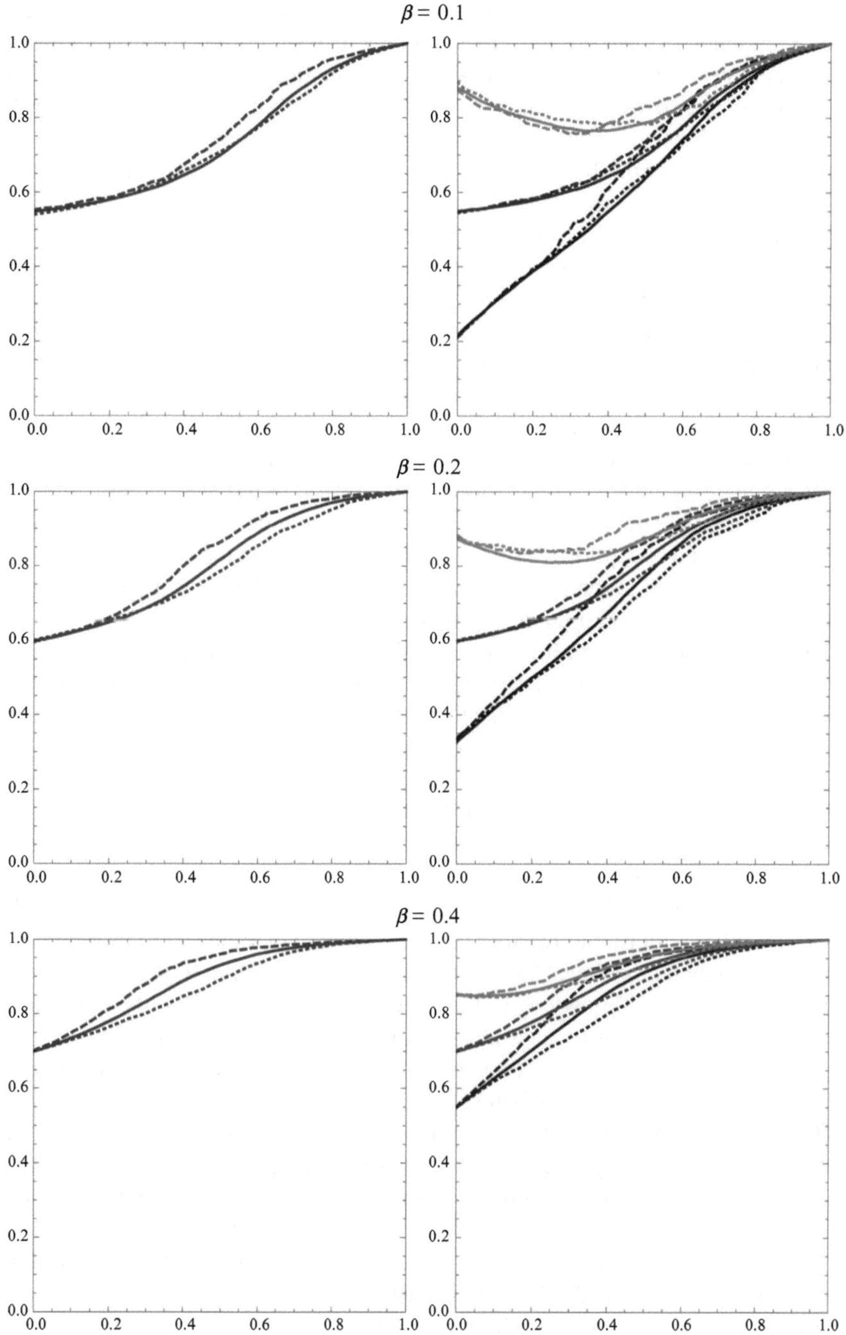

Fig. 5.3 Ensemble-wide mean agreement evolutions averaged over all position pairs (*left-hand panels*) and averaged over position pairs with specific initial agreement (*right-hand panels*). Initial agreement intervals as in Fig. 5.1. Curves represent means of all debates (*solid*), compact debates (*dotted*), and fragmented debates (*dashed*)

Apparently, the results of the previous chapter are turned upside down. While the general tendency of background knowledge to foster rapprochement is plain both in compact and in fragmented debates; the introduction of background knowledge causes proponent positions in fragmented debates to approach each other faster than in compact debates—as opposed to a comparatively more rapid rapprochement in compact debates without background knowledge. Thus, whereas the dashed curves in Fig. 4.6, representing the average over all fragmented debates, lie well below the corresponding dotted curves, this is not the case in Fig. 5.3 anymore. The broader the background knowledge, the faster the rapprochement of proponent positions in fragmented debates as compared to compact debates: With $\beta = 0.1$, ensemble-wide mean agreement in fragmented debates (dashed curves, left-hand panels) outgrows mean agreement in compact debates (dotted curves, left-hand panels) only at a density of 0.4. With $\beta = 0.4$, however, fragmented debates exhibit more substantial agreement incline right from the beginning. At a density of 0.4, mean agreement in fragmented debates lies almost 10 percentage points above mean agreement in compact debates. Considering positions with specific initial agreement (right-hand panels) confirms the general finding that mean agreement rises faster in fragmented debates—with one exception. For narrow background knowledge, $\beta = 0.1$, position pairs with high initial agreement in compact debates display, at densities below 0.4, slightly higher agreement than in fragmented debates.

Background knowledge affects compact and fragmented debates differently in terms of the average number of nonidentical positions, as well. The left-hand plots in Fig. 5.4 describe how the number of nonidentical positions declines as inferential density increases. Not only does the ensemble-wide average over all debates (solid curves) decline much more rapidly as compared to a *random argumentation* without background knowledge (see Fig. 4.3), but it also starts to decline earlier: without background knowledge, the number of non-identical positions remains unchanged for inferential densities below 0.5, while in debates with background knowledge, this number has dropped from 6 to 5.5 at $D = 0.5$ ($\beta = 0.1$), and even to roughly 3 ($\beta = 0.4$). What's more, the number of nonidentical positions declines much more quickly in fragmented debates than in compact debates. So, in a fragmented debate with 40% background knowledge, there remain, on average, slightly more than 3 distinct proponent positions at a density of 0.4, as opposed to ca. five different proponent positions in a compact debate. Proponents are much more likely to reach full agreement in fragmented debates.

The collapse densities, that is, the inferential densities at which two proponent positions reach, on average, full agreement, are displayed in the right-hand plots of Fig. 5.4. They corroborate the general picture. Even with a background knowledge comprising 10% of the sentences, proponent positions tend to collapse at lower densities in case the space of coherent positions is fragmented rather than compact. This difference (between the bars with dashed and dotted border) becomes more pronounced as background knowledge increases. For example, with $\beta = 0.4$, two positions which initially agreed by more than 80% of the sentences reach, in a fragmented debate, full agreement at a density of 0.42. This compares with an average collapse density of 0.55 in compact debates.

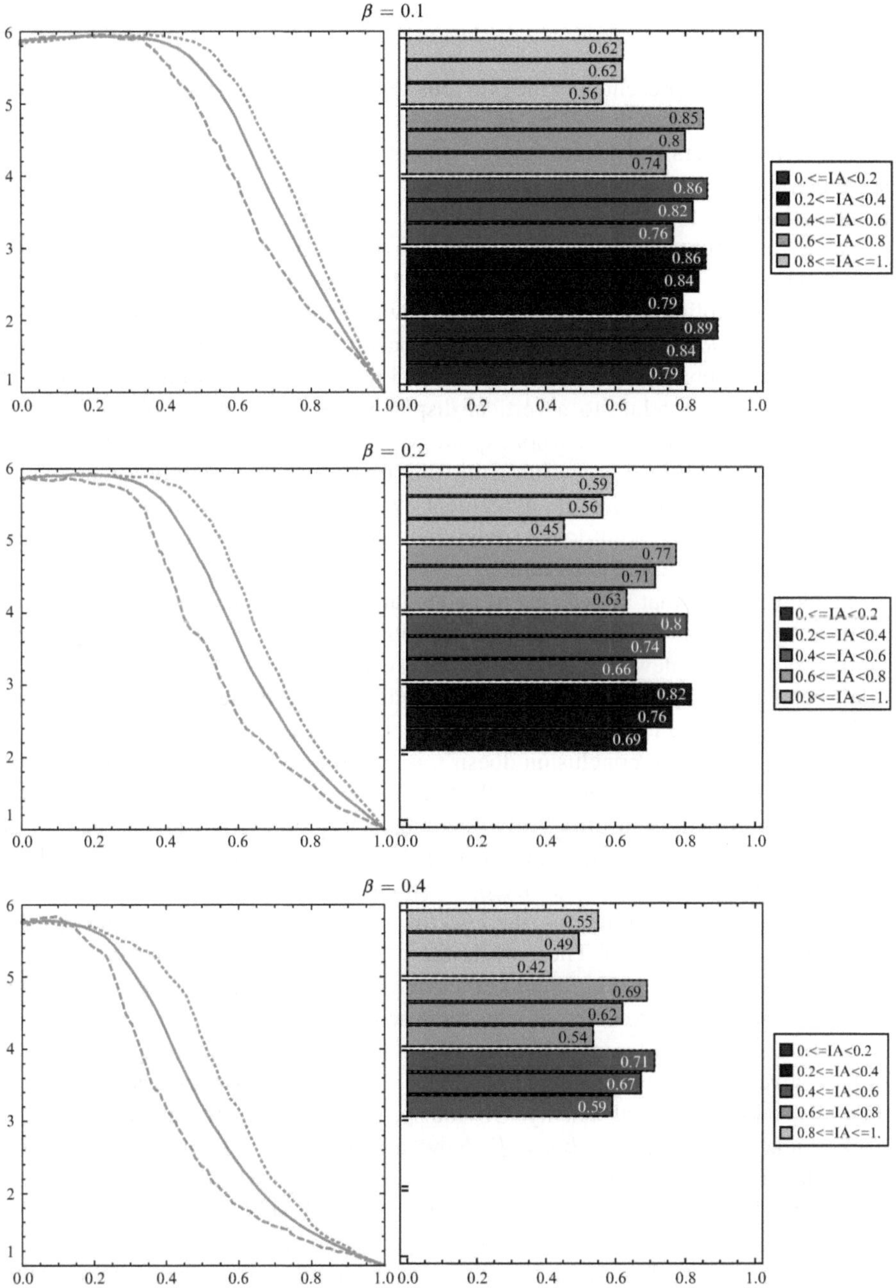

Fig. 5.4 *Left-hand panels*: ensemble-wide mean number of non-identical positions. *Right-hand panels*: ensemble-wide mean collapse inferential densities for position pairs with specific initial agreement. Values reflect means over all debates (*solid curves* and *borders*), compact debates (*dotted*), and fragmented debates (*dashed*)

5.3 Discussion

The various results presented in the previous section reveal two main features which are in need of explanation. This is, firstly, the fact that background knowledge substantially fosters the mean rapprochement between proponent positions, and, secondly, the greater consensus-conduciveness of argumentation in fragmented as compared to compact debates given a body of shared background beliefs.

As a first thing to note, the increased rapprochement due to background knowledge does not merely consist in the fact that the proponents agree with respect to the explicitly fixed background knowledge, to which we shall refer as the *basic background knowledge.* As the basic background knowledge stays constant throughout the debate, this fact merely results in a uniform raise of mean agreement (by $\beta/2$)—corresponding to a vertical displacement of the ensemble-wide mean agreement curve—and can thus not explain why proponents reach higher agreement *relative to their initial agreement.* Still, this first effect is certainly a part of the larger story which explains the agreement evolution in debates with background knowledge. In particular, it does explain the elevated initial mean agreement in the absence of any inferential constraints.

The key insight that allows one to understand the position dynamics in debates with background knowledge consists in the following observation. As the argumentation proceeds and new inferential relations are discovered, additional theses which don't belong to the basic background knowledge can nevertheless be derived from it. If, for example, an argument whose premisses belong to the basic background knowledge, but whose conclusion doesn't, is introduced, this conclusion has to be agreed upon by all proponents, too. Without being explicitly included in the body of background beliefs, it effectively becomes background knowledge because of the inferential relations that hold between the debate's sentences. Let us call all sentences whose truth values are fixed, given the basic background knowledge and the debate's inferential relations, the *effective background knowledge* (and β_{eff} the corresponding proportion). Thus, while the basic background knowledge remains constant, the effective background knowledge increases in the course of a debate, and forces proponents to agree on ever more sentences.

Our definition of inferential density allows us to derive a simple approximation of how, given a basic background knowledge, the effective background knowledge increases with inferential density. Obviously, if $D(\tau) = 0$, effective equals basic background knowledge and $\beta_{\text{eff}} = \beta$. Now, assume $D(\tau) > 0$. Inferential density relates the number of binary decisions a proponent makes when positioning herself in a debate to the number of sentence pairs in the debate's sentence pool (n). We may equate the number of binary choices with the number of sentences whose truth values a proponent is free to set, as opposed to the number of sentences whose truth values are, subsequently, automatically determined by the inferential relations that make up the dialectical structure. So, for example, if $D(\tau) = 0.5$, a proponent is free to set the truth values of $n/2$ sentences; the remaining ones are fixed automatically given the inferential constraints. If $D(\tau) = 1/4$, a proponent may fix 3/4 of all

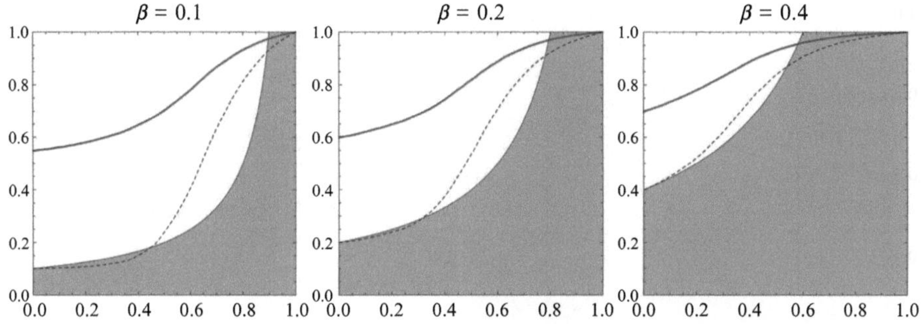

Fig. 5.5 Ensemble-wide mean agreement evolutions averaged over all position pairs (*thick curves*) and proportion of effective background knowledge, β_{eff}—both, ensemble-wide mean (*dashed curve*) and as approximated by Eq. (5.1) (*shaded area*)—plotted against inferential density

the sentences; one quarter of the truth values is, accordingly, determined by the arguments. Generally, specifying a proportion of $1 - D(\tau)$ of all the n sentences determines the remaining truth values. Here comes our approximation: We shall assume that not only fixing $(1 - D(\tau)) \cdot n$ sentences determines the truth values of the remaining $D(\tau) \cdot n$ ones but setting any number r of truth values ($r \leq (1 - D(\tau)) \cdot n$) implies that a corresponding number r^+ of further truth values is determined by the inferential relations, with $r/r^+ = (1 - D(\tau))/D(\tau)$. So, let us assume that the basic background knowledge consists of r sentences. The corresponding effective background knowledge comprises, consequently, $r + r^+$ sentences, and we have,

$$
\begin{aligned}
r^+ &= \frac{D(\tau)}{1 - D(\tau)} r \\
r + r^+ &= \left(1 + \frac{D(\tau)}{1 - D(\tau)}\right) r \\
\frac{r + r^+}{n} &= \left(1 + \frac{D(\tau)}{1 - D(\tau)}\right) \frac{r}{n} \\
\beta_{\text{eff}} &= \left(1 + \frac{D(\tau)}{1 - D(\tau)}\right) \beta. \qquad (5.1)
\end{aligned}
$$

Equation (5.1) approximates the functional relation between effective background knowledge, basic background knowledge, and inferential density. The ratio of effective background knowledge, β_{eff}, represents a lower boundary to the normalized agreement for any two position pairs, and, consequently, for debate- as well as ensemble-wide mean agreement. It is plotted, together with ensemble-wide mean agreement evolutions, in Fig. 5.5. As this figure demonstrates, the analytic approximation provides a fair estimate at high and low densities yet tends to underestimate the ratio of effective background knowledge at medium densities. In the subsequent calculations, we will therefore rather rely on the ensemble-wide mean ratio of effective background knowledge as derived from the simulation data.

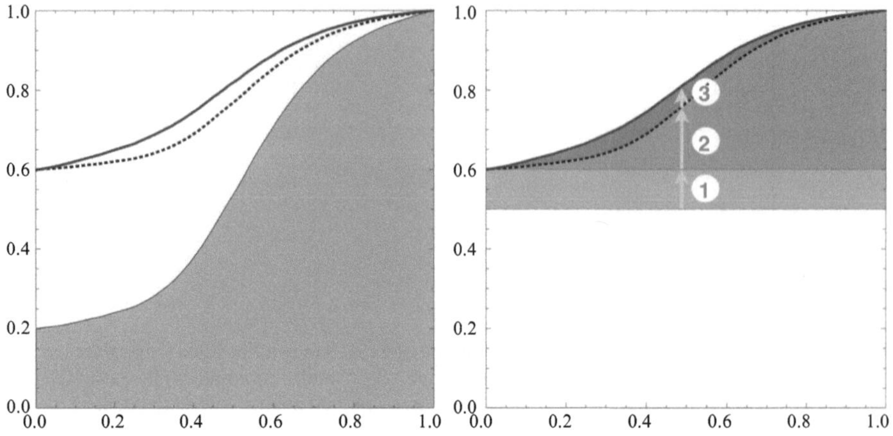

Fig. 5.6 *Left-hand panel*: expected agreement of randomly assigned positions which merely coincide with respect to the effective background knowledge (*dashed*). The ensemble-wide mean ratio of effective background knowledge for $\beta = 0.2$ is plotted as *gray area*, the ensemble-wide mean agreement as solid curve. *Right-hand panel*: illustration of the three mechanism which lead to rapprochement of proponent-positions in debates with background knowledge. (1) Agreement with regard to basic background knowledge. (2) Agreement with regard to effective background knowledge. (3) Agreement due to the contraction of the space of coherent positions

Consider two randomly assigned and not necessarily dialectically coherent positions which agree with regard to the effective background knowledge (β_{eff}). As the sentences which don't belong to the effective background knowledge take random truth values, the two positions exhibit, on average, a normalized agreement equal to $(1 + \beta_{\text{eff}})/2$. So this formula describes agreement evolution, plotted as dotted curve in the left-hand panel of Fig. 5.6, we'd expect to observe if proponent positions took the effective background knowledge into account and evolved, otherwise, randomly. We may thus identify the second mechanism that generates the superior rapprochement in random debates with background knowledge: Effective background knowledge raises the expected mean agreement of randomly assigned positions far above the expected value given the basic background knowledge.

A third and final mechanism that rounds up our explanation of the higher effectiveness of an argumentation with background knowledge simply consists in the effect of inferential constraints and the resulting deformation of the space of coherent positions. The contraction of the SCP causes the mean mutual agreement to rise well above the mean agreement of random (and not necessarily coherent) positions which merely share the effective background knowledge. The right-hand panel of Fig. 5.6 illustrates these three mechanisms which bring about the substantially faster rapprochement in random debates with background knowledge. As this figure demonstrates, the second mechanism, that is, the expansion of the effective background knowledge, is the dominant driver of the rapid rapprochement.

We shall now turn to the second peculiar result reported in the previous section: the relatively fast agreement in fragmented as compared to compact debates. This turns the findings regarding random debates without background knowledge upside down (see Sect. 4.4). Contrary to one's first impression, however, these results don't contradict our previous reasoning. On the opposite, our hypothesis about the two types of debate dynamics yields an explanation for the prima facie surprising results, and is thus confirmed. In a nutshell, common background knowledge tends to force proponent positions on one and the same segment of a fragmented space of coherent positions. As such fragments contract faster than the entire space of coherent positions, proponent positions approach each other relatively rapidly. In the absence of any fragmentation (in a compact debate, i.e.), background knowledge merely pushes the proponent positions into the same region of the well-connected, compact space of coherent positions. In that case, proponent positions will approach each other no faster than the entire space of coherent position contracts.

The metaphors developed in the previous chapter may serve to illustrate these different dynamics. Consider the fishing-net metaphor, representing the compact evolution of a space of coherent positions, first. In this analogy, the introduction of background knowledge corresponds to the fact that, for whatever reason, the fishes all assemble in one section, say one half, of the fishing net, leaving the remaining volume completely deserted. The proportion of the entire net's volume which is occupied by the fishes declines as the net is pulled (expansion of effective background knowledge), leaving the fishes gathered in an ever tinier segment of the fishing net. The fishes won't be packed in virtually one and the same place before the net is entirely pulled. However, this agreement will still be reached in a continuous, steadily increasing way.

The position dynamics in a fragmented debate is illustrated by the flooded-village analogy. Common background knowledge may be represented, in this case, by the assumption that a certain area of the village is completely void of inhabitants, no matter whether the buildings in that area are already entirely flooded or not. To begin with, the fact that the villagers are initially (and not merely coincidentally) located close to each other prevents that flooding will drive them on very distant buildings. Instead, the rising watermark will cause the inhabitants to climb closely related buildings, or, depending on the number of different buildings in that village, to mount the very same structure. Even a very modest flooding may thus compel the villagers to adopt a common position, completely suppressing their freedom to move, if, by chance, they are forced to gather in a part of the village where there is only one elevated building nearby. Should the inhabitants, however, search shelter on different roofs in the initial phase of the flooding, the effective background knowledge—causing villagers to abandon ever larger parts of the village irrespective of the flooding—will bring about the roof hopping much more quickly and cause positions to approach each other in steep steps.

We have suggested in the previous chapter that there is a trade-off between generating group cohesion and overall agreement. While mean agreement grows steadily in compact debates, fragmented debates give rise to different groups of proponents who internally agree to a large extent—with adverse effects for overall agreement. This opinion-island effect provides a further conceptual framework to describe and understand the dynamics with background knowledge, for background knowledge ensures that many isolated opinion-islands are inhabitable in the very first place, causing proponents to form closely related, or a single, opinion group. That's why, with background knowledge, fragmentation of the space of coherent positions and the resulting group cohesion does not undermine overall agreement at all. The trade-off seems to vanish.

So, these different dynamics do not only explain why, with larger background knowledge, mean agreement rises more rapidly in fragmented debates but also render the relatively fast collapse of different proponent positions in fragmented debates with background knowledge intelligible.

Chapter 6
Comparing the Consensual Dynamics of Four Proponent-Specific Argumentation Strategies in Dualistic Debates

At the end of Chap. 4, we suggested two levers for increasing the agreement which is brought about by rational argumentation, namely (1) introducing background knowledge, and (2) devising argumentation strategies which cause proponent positions to converge substantially at densities well below 0.5. In the previous chapter, we pursed option (1) and showed how commonly shared background beliefs increase mean agreement. This chapter, as well as the following ones, focuses on the second alternative. More precisely, we will set up four different rules according to which proponents introduce new arguments into the debate. These rules can be understood as argumentation strategies adopted by the proponents. They replace the simple *random argumentation* mechanism the previous simulations relied upon.

The four argumentation strategies we will consider can be developed along the following lines. Consider a proponent who is about to introduce an argument into the debate. She is aware of her own position, her opponents' positions, the shared background knowledge (which we assume, throughout this chapter, to be empty), and the current dialectical structure. The construction of the new argument requires that two questions be addressed: (a) Which sentence is supposed to be the conclusion of the new argument? (b) What are its premisses? Each of these questions corresponds to a basic choice the proponent faces. With respect to (a), the proponent may decide to argue in favor of her own position (a.1), or she may want to argue against the position of (at least) one of her opponents (a.2). As regards (b), she may base her new argument on premisses she considers true (b.1), or, alternatively, she develops her argument on the basis of sentences which one of her opponents accepts (b.2). This results in a matrix of four alternative argumentation strategies which we shall name "*fortify*," "*convert*," "*attack*," and "*undercut*" (see Table 6.1). A proponent *fortifies* her own position, if she puts forward an argument containing but premisses and a conclusion she believes to be true. In the other three cases, the proponent considers, besides her own position, the stance of a randomly chosen opponent. She tries to *convert* the opponent if her argument (1) rests on premisses the opponent agrees with and (2) backs a conclusion adopted by the proponent. Arguing, based on premisses the proponent adheres to, against

G. Betz, *Debate Dynamics: How Controversy Improves Our Beliefs*,
Synthese Library 357, DOI 10.1007/978-94-007-4599-5_6,

Table 6.1 Four argumentation strategies a proponent with position $\mathscr{P}$ may adopt when designing a new argument by choosing (a) its conclusion c and (b) its premisses p_1, p_2, where $\mathscr{Q}$ is the position of the opponent addressed by the argument

	(b.1) $\mathscr{P}(p_1) = \mathscr{P}(p_2) = \text{true}$	**(b.2)** $\mathscr{Q}(p_1) = \mathscr{Q}(p_2) = \text{true}$
(a.1) $\mathscr{P}(c) = \text{true}$	*fortify*	*convert*
(a.2) $\mathscr{Q}(c) = \text{false}$ and $\mathscr{P}(c) \neq \mathscr{Q}(c)$	*attack*	*undercut*

the position of the opponent is referred to as *attacking* the opponent's position. If the new argument, finally, takes off from the opponent's position in order to demonstrate that a further conviction of the opponent, which does not figure as a premiss in this argument, is false, the opponent's position is *undercut*. The last two strategies presume, moreover, that opponent and proponent disagree with regard to the argument's conclusion (this ensures that a proponent doesn't undermine her own position). As a consequence, every new argument which conforms with the *attack* rule represents eo ipso a *fortification* of the proponent's position, and an argument which *undercuts* an opponent complies with the *convert* strategy, as well.

It goes without saying that these four argumentation strategies by no means exhaust the spectrum of argumentation strategies proponents may possibly adopt. They represent, however, very simple and ideal types of defensive and offensive argumentation.

In this chapter, we describe and discuss simulations of debates where the four argumentation strategies compete against each other.

6.1 Setup

For each (unordered) pair of argumentation strategies, that is, *fortify–fortify, fortify–convert, fortify–attack*, ..., an ensemble of at least 2,000 debates is set up in line with the following mechanisms. This gives rise to ten different ensembles, studied in this chapter. Each debate unfolds over a pool of $2 \cdot 20$ sentences and comprises two proponents.

Argumentation Mechanism: The ensemble-specific pair of argumentation strategies, for example, *convert–attack*, defines the argumentation rules followed by the two proponents in each debate. One proponent implements the first strategy, her opponent the second. In alternating sequence, the proponents put forward new arguments—one per step—in accordance with their corresponding argumentation strategy.

Discovery Mechanism: The background knowledge $\mathscr{B}$ is empty.

Update Mechanism: *Closest coherent* (cf. Sect. 4.1).

A debate simulation stops, if the two proponents reach full agreement, or if the inferential density has increased beyond 0.8.[1]

By simulating but two proponents per debate, we deviate—seemingly unnecessarily—from the design of our previous simulations. The reduction of the number of proponents per debate results, however, from the following reasoning. In contrast to debates with random argument construction, the number of proponents per debate crucially influences the simulation results when arguments serve proponent-relative purposes. This is because the more proponents there are, the less frequently a new argument will directly address—that is, be constructed with explicit consideration of—a particular proponent position. Thus, the more proponents engage in a debate, the more the argumentation—despite unfolding in accordance with some of the four argumentation strategies—resembles, in general, a random argumentation. In a controversial dialogue between two proponents, however, at least every second argument put forward directly addresses a proponent's position. As we aim at studying the effects of argumentation strategies, it is prudent to set up the simulations so as to magnify the effect of the different strategies. As the number of proponents does, however, not affect the outcome of debates with *random argument construction*, the random debates studied in the previous chapters may still serve as a benchmark when investigating the two-proponent debates. Since, conversely, the results from the two-proponent debates cannot be scaled to debates with many proponents that implement the respective argumentation strategies, we will explicitly investigate, in the next chapter, debate simulations which (1) rely on a sophisticated, multi-proponent argumentation mechanism derived from the four basic argumentation strategies and which (2) comprise, in the same time, six proponents.

6.2 Results

The mean agreement evolution for each of the ten ensembles is given in Fig. 6.1. As a first thing to note, the agreement evolutions vary significantly. It is plain, at first glance, that the argumentation strategies exert a major influence on the position dynamics. Let us investigate the various results step by step. First, consider the three plots in the upper half of Fig. 6.1. They display the agreement evolution of

[1] Since proponent positions will eventually collapse onto the one remaining coherent position at a density of 1, the final steps towards reaching that density are of no particular interest. Yet even more importantly, in some specific situations, there exists no potential new argument which may force the proponents to modify their positions. For example, if the two proponents agree with regard to one single sentence only, no argument that satisfies the *attack* rule renders any proponent position incoherent. In these cases, simulations risk to continue ad infinitum. Experience has shown that simulations with the attack rule are particularly prone to this threat. To alleviate the problem, we have stipulated that (1) a random argument shall be introduced if there is no potential argument whatsoever which satisfies the corresponding argumentation rule and (2) simulations abort if a density of 0.8 has been reached.

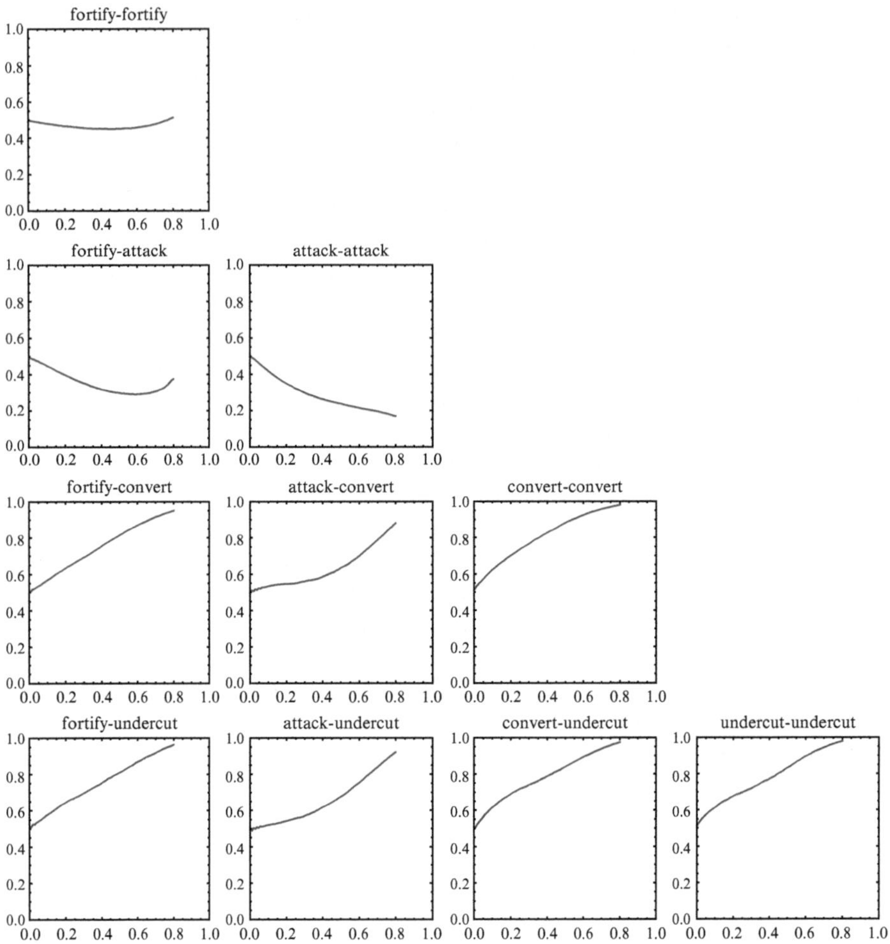

Fig. 6.1 Ensemble-wide mean agreement evolutions for ten ensembles with two proponents each who pursue the argumentation strategies indicated on top of the diagrams. Mean agreement is plotted against inferential density

debates where proponents follow the *attack* or the *fortify* rule. Strikingly, agreement does, on average, not increase in these debates. It either stays roughly the same (*fortify–fortify*) or decreases significantly. In the ensemble *attack–attack*, mean agreement falls to less than 0.2 at a density of $D = 0.8$. Hence, proponents are systematically driven apart. Next, we shall have a closer look at the three ensembles presented in the right half of our figure, that is, ensembles where proponents argue in accordance with the *undercut* or the *convert* rule. These debates exhibit a remarkable rapprochement—way above the random debates we have previously studied. At a density of 0.8, agreement has virtually reached, in all three ensembles, 100%. The speed of this convergence, however, varies slightly. In the ensemble *convert–convert*, mean agreement increases relatively rapidly until a density of 0.4 and slows down subsequently. The ensemble with two proponents who follow the *undercut* rule, in contrast, gives rise to a relatively constant increase of mean agreement. Finally, there are the four ensembles where a *fortify* or *attack* strategy on the one side meets a *convert* or *undercut* strategy on the other side. Of these, the ensembles with a fortify rule (*fortify–convert,fortify–undercut*) display very similar, constant rapprochements, which result in a final mean agreement that is almost as high as in the *undercut–undercut* case. The remaining two ensembles (*attack–convert, attack–undercut*) show a mean agreement evolution which resembles, regarding its shape, the dynamics of random debates. The agreement increases only slightly at low densities, before rapprochement accelerates once a certain inferential density has been passed. Compared to the other ensembles in these two rows, *attack–convert* and *attack–undercut* reach roughly similar levels of agreement at high densities but display considerably smaller agreement at lower densities.

So far, we have considered mean agreement averaged over all position pairs in a debate. This doesn't reveal how the rapprochement of proponent positions depends on their initial agreement—and that is what we will consider next. Figure 6.2 plots the familiar mean agreement evolutions of different groups of position pairs (namely, with high, medium, and low initial agreement) for each of the ten ensembles. Unsurprisingly, these evolutions vary to a large extent, too. We shall, again, consider the ensembles step by step. In the three ensembles displayed in the upper half of Fig. 6.2 (with *attack* and *fortify* only), mean agreement increases in no proponent group whatsoever. Even the proponent positions with very low initial agreement (bottom curves) don't approach each other. The agreement level regarding positions with medium or high initial agreement either stays roughly the same (*fortify–fortify*) or decreases more or less dramatically. So, in the *attack–attack* ensemble, mean agreement of positions with high initial agreement drops by 50 percentage points from 0.9 to 0.4! These results contrast starkly with the dynamics of the three ensembles displayed at the right-hand side of the diagram. Here, dovetailing with the general findings of Fig. 6.1, the mean agreement tends to increase for the different groups of proponents. We may, however, discern a notable, qualitative difference in the position dynamics. Consider the ensembles *convert–convert* and *undercut–undercut*. Whereas positions with high initial agreement virtually don't depart from each other in the first case, there is a significant drop in mean agreement of initially very close positions in the latter one. This drop coincides with a comparatively rapid rapprochement of proponent positions with very low initial agreement: The bottom curve catches up with the middle one at $D = 0.2$, corresponding to an agreement increase by 55% (compared to 45% in the ensemble *convert–convert*). Apropos of the remaining four ensembles, they fall, again, into two similar pairs. The two ensembles with a *fortify* strategy exhibit comparatively strong agreement increase with respect to the three proponent groups. Interestingly, however, mean agreement of closely related initial positions drops considerably in *fortify–undercut*, yet doesn't in *fortify–convert*. The two ensembles *attack–convert* and *attack–undercut* are, finally, characterized by a drastic decline of mean agreement for positions with high initial agreement. In both ensembles,

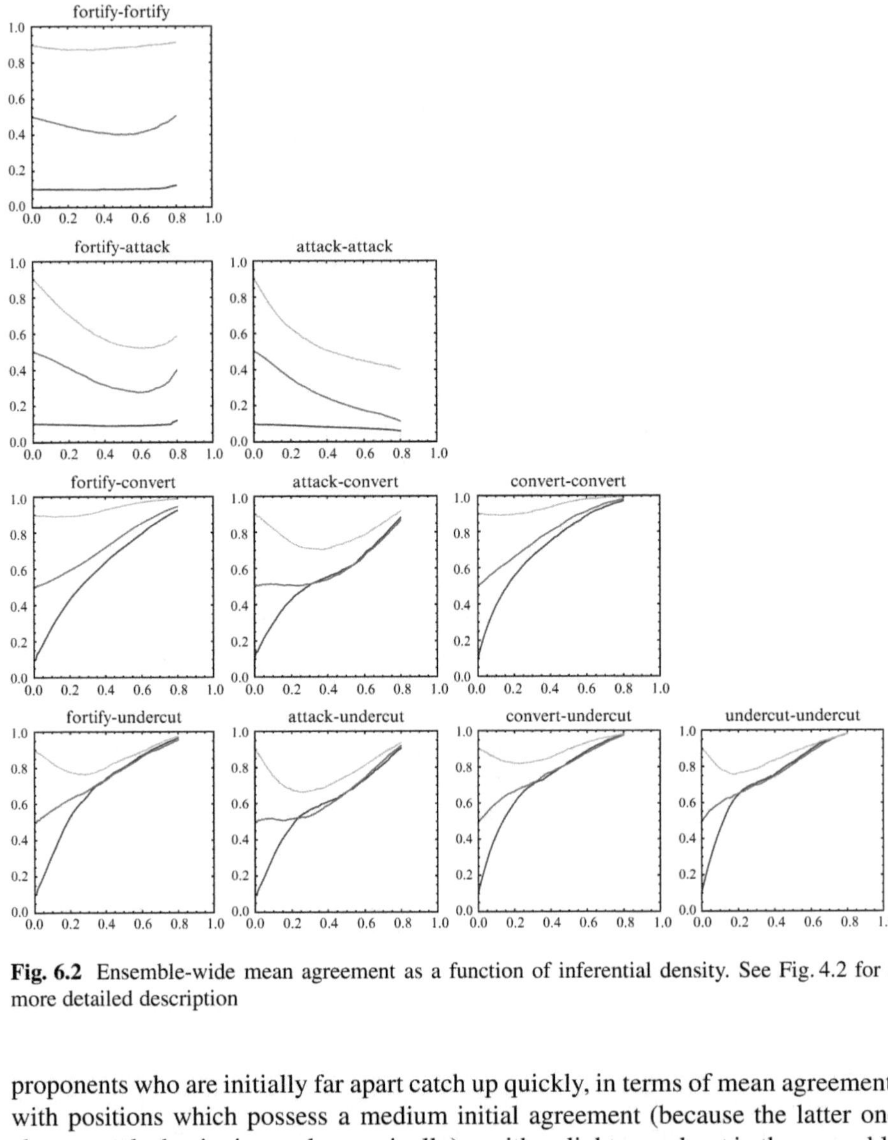

Fig. 6.2 Ensemble-wide mean agreement as a function of inferential density. See Fig. 4.2 for a more detailed description

proponents who are initially far apart catch up quickly, in terms of mean agreement, with positions which possess a medium initial agreement (because the latter one change, at the beginning, only marginally)—with a slight overshoot in the ensemble *attack–undercut*.

The evolution of the average number of nonidentical proponent positions, plotted in Fig. 6.3, provides a third perspective on our ensembles. With two proponents per debate, this number varies, obviously, between 1 (full consensus in all the debates in the ensemble) and 2 (no full agreement in any debate). Initially, the mean number of nonidentical positions is slightly smaller than 2. This is because, due to the random assignment of proponent positions, some positions initially agree by coincidence. In the ensembles with *fortify* and *attack* only (upper three plots),

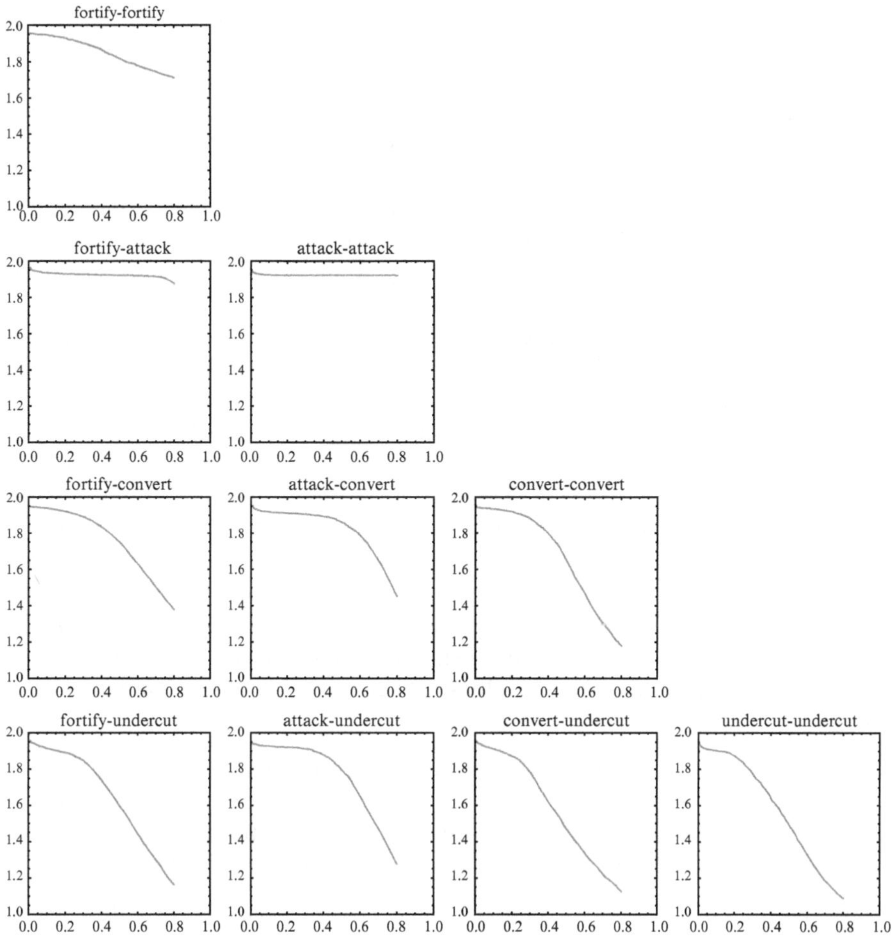

Fig. 6.3 Ensemble-wide average number of nonidentical positions in the ten ensembles, plotted as a function of inferential density

proponent positions reach full agreement hardly at all. With the average number of nonidentical positions falling to 1.7 in *fortify–fortify*, consensus emerges in 30% of the debates at a density of 0.8. In the other two ensembles (*fortify–attack*, *attack–attack*), proponents virtually never reach full consensus. Like in the previous figures we have studied, the ensembles shown on the right-hand side of Fig. 6.3 display a completely different dynamic. Full agreement is reached, at a density of 0.8, in more than 80% of the debates, in the case of *undercut–undercut* in even more than 90%. The ensemble-wide mean number of nonidentical positions declines much more rapidly than in a purely random debate (cf. Fig. 4.3). In contrast to the mean agreement evolutions studied above, the *undercut* rule outperforms the *convert* strategy in terms of engineering full consensus: While, in the *undercut–undercut* ensemble, half of the debates exhibit full agreement at a density of 0.5, this holds for less than 40% in the *convert–convert* ensemble. The remaining four ensembles (*fortify* or *attack* on the one side, *convert* or *undercut* on the other side) give rise to mixed pictures. We observe a notable decline in the average number of nonidentical positions in the two ensembles with a proponent who fortifies her position. In the ensembles *attack–convert* and *attack–undercut*, however, the number of non-identical positions hardly changes during the initial phase and drops steeply at higher densities. This harmonizes with our previous results. As far as these four ensembles are concerned, the *undercut* strategy, again, seems to be more successful than the *convert* rule in bringing about complete consensus.

The findings presented in Fig. 6.3 suggest that *convert* and *undercut* are, in sum, much more suited for reaching consensus than *fortify* or *attack*. More specifically, *undercut* seems to be more consensus-conducive than *convert*, and *fortify* more so than *attack*. Now, it is one thing to ask if and how rapidly proponents do attain a consensus, and it is another thing to ask what kind of consensus is eventually reached. So, for example, how far do proponents have to depart from their initial position in order to concur with their opponents? Are there, in particular, argumentation strategies which allow a proponent to stick to her position while reaching a consensus nevertheless? Do, moreover, argumentation rules which favor consensus cause the proponents to alter their own positions to a larger extent? In order to answer these questions, we consider for each *argumentation context* **X**Y—where an argumentation context is fully specified by the proponent's argumentation rule (**X**) and the opponent's one (Y):

1. The collapse density, that is, the inferential density at which full consensus is typically reached in a debate where the respective strategies are implemented.[2]
2. The "versatility" of the proponent position, measured as the distance between the final and the initial position of the proponent (who follows the rule **X**).

Figure 6.4 plots for each argumentation context the proponent's versatility against the corresponding collapse density.

What does this plot tell us? In order to understand it, we shall go through it step by step, reidentifying the groups of ensembles we have studied in the previous figures. Consider the debates where proponents apply the *attack* or the *fortify* strategy only. They correspond, in Fig. 6.4, to the argumentation contexts **F**F, **F**A, and so on. These items are situated in the lower right corner of the diagram. This means (a) in these debates, consensus emerges, as we already know, rather late, if at all; and (b) the proponent positions in such argumentation contexts are not very versatile, that is, change only slightly in the course of a debate. More precisely, however, a proponent in the argumentation context **F**A (i.e., a proponent who *fortifies* her position and faces an opponent that follows the *attack* rule) is more versatile than in the contexts **F**F, **A**A, and, in particular, **A**F. That is, a proponent who *fortifies* her position in the *fortify–attack* ensemble alters, during the debate, 35% of her convictions

[2]If full consensus isn't reached at a density below 0.8, we posit a collapse density of 1.

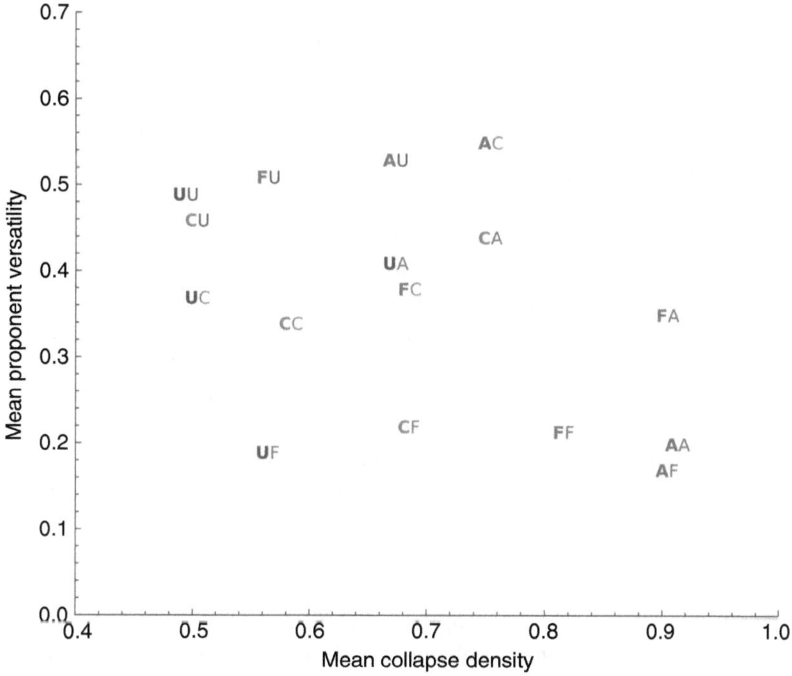

Fig. 6.4 Versatility and mean collapse density of the four argumentation strategies. For each argumentation context **X**Y (cf. text), the ensemble-wide mean distance between the initial and final position of proponents who implement strategy X, as derived from the X–Y ensemble, is plotted against the mean collapse density of positions in the X–Y ensemble. Argumentation rules are abbreviated by their first letters

(not counting modifications that are undone), whereas the proponent with the *attack* strategy merely disagrees, in the end, with 17% of her initial beliefs. We shall say, to describe such a fact concisely: The *attack* rule "dominates" the *fortify* rule. Let us, next, consider the ensembles where proponents apply either the *convert* or the *undercut* strategy. The corresponding argumentation contexts are located in the upper left part of the diagram. They represent high versatility/high consensus-conduciveness settings. Full agreement emerges in these contexts typically at densities between 0.5 (**U**U) and 0.59 (**C**C), and proponents modify, on average, 34–49% of their initial convictions. Moreover, *undercut* dominates *convert* since the versatility of **U**C is lower than of **C**U. Finally, we have the four ensembles where *attack* or *fortify* on the one side faces *undercut* or *convert* on the other side. Consider, first, the two ensembles with *fortify*, corresponding to four argumentation contexts: **F**C, **C**F, **F**U, **U**F. Consensus emerges in these contexts, on average, below a density of 0.7—which signals still a relatively high consensus-conduciveness. In terms of versatility, however, the proponents perform very differently: While an **F**U proponent typically deviates by more than 50% from her initial position, her opponent, represented by the **U**F context, modifies less than 20% of her original beliefs. Because **FC** exhibits higher versatility than **CF**, *convert* dominates *fortify*, too. Eventually, we shall examine the remaining two ensembles with the *attack* strategy. There are four argumentation contexts: **AU**, **UA**, **AC**, **CA**. These contexts display medium consensus-conduciveness (mean collapse density of around 0.7), but a high versatility. This holds in particular for proponents who implement the *attack* strategy while being opposed by the *convert* or the *undercut* rule. They disagree, at the end of the debate, with more than half of their initial beliefs. Although the proponents in the contexts **UA** and **CA** modify clearly fewer of their original convictions (*attack* is dominated both by *convert* and by *undercut*), their final positions deviate significantly from their initial ones, too.

The dominance relation we can extract from Fig. 6.4 is a transitive and asymmetric relation such that *undercut* dominates *convert*, *convert* dominates *attack*, and *attack* dominates *fortify*. This means, for example, that a proponent who follows the *undercut* rule will deviate, on average, no more from her initial position than her opponent, no matter which strategy her opponent implements.

The versatility of proponents in an argumentation context is determined by the proponent's strategy as much as by the opponent's one. Remarkably, as Fig. 6.4 shows, proponents who oppose an undercut strategy possess a very high versatility—quite irrespective of the rule they follow themselves. Likewise, the argumentation contexts where the opponent fortifies her position possess the least versatile proponent positions. Confronted with a fortify strategy, proponents can apparently stick to their initial point of view, forcing, possibly, their opponent to move.

6.3 Discussion

In this section, we will discuss, and try to explain, the following stylized facts that can be extracted from the various results presented above:

1. In general, the four argumentation strategies seem to fall, roughly, into two pairs which give rise to similar outcomes: *fortify*/*attack* on the one hand and *convert*/*undercut* on the other hand.
2. Regarding mean agreement, averaged over all position pairs in an ensemble:
   (a) The strategies *convert*/*undercut* foster rapprochement to a significantly larger extent than *fortify*/*attack*.
   (b) The *attack* rule performs, in terms of mean agreement increase, worse than the *fortify* rule.
   (c) The *convert* rule leads to even more rapid agreement increase than the *undercut* rule.
3. Regarding mean agreement of proponent groups with specific initial agreement:
   (a) The *attack* rule tends to undo high initial agreement, whereas the *fortify* rule doesn't.

   (b) Neither *fortify* nor *attack* increases the agreement between initially very distant positions.
   (c) The *convert* rule doesn't cause initially high agreement to vanish—while the *undercut* rule does. The decline of high agreement, with *undercut*, coincides with a rapid rapprochement of very distant positions.

4. Regarding the number of nonidentical positions:

   (a) In debates with *fortify/attack* only, full consensus is hardly ever reached at all. The number of nonidentical positions decreases, however, quickly with *convert/undercut*.
   (b) In terms of bringing about full agreement, *fortify* is, nevertheless, somewhat more effective than *attack*.
   (c) The *undercut* rule appears to be slightly more consensus-conducive than the *convert* rule.

5. Regarding the versatility of proponent positions:

   (a) The strategies *convert/undercut* give rise to argumentation contexts with high versatility and high consensus-conduciveness; *fortify/attack*, in contrast, display low versatility as well as low consensus-conduciveness.
   (b) Based on a pairwise comparison of versatility, *undercut* dominates *convert*, *convert* dominates *attack*, and *attack* dominates *fortify*.
   (c) Proponent positions possess a high (low) versatility in argumentation contexts where they are opposed by the *undercut* rule (*fortify* rule).

The first item in this list of stylized facts is not difficult to explain. As we have noted when introducing the four strategies, every argument which is put forward in line with the *attack* rule necessarily conforms with the *fortify* rule, as well. Likewise, an argument that *undercuts* a position *converts* it in the same time. These relationships, which logically follow from the definitions of the four strategies, explain the close resemblance we observe in our simulation results.

In order to understand the other stylized facts enumerated above, we have to obtain a more precise conceptual understanding of how the argumentation strategies operate. This can be achieved by studying how arguments, put forward in accordance with one of the rules, shape the space of coherent positions. That is what we will do next. Having analyzed the effects of the different strategies on a theoretical level, we will, later in this section, return to the stylized facts which are to be explained.

We study, in the following conceptual investigation, proponent positions which are defined on three different sentences and which are, as a consequence, located in a comparatively small space of coherent positions—made up, initially, of eight positions. This space of coherent positions can be visualized by a cube, each edge of the cube representing a truth value assignment to the three sentences. In our visualizations, the x-dimension corresponds to the truth value of the first sentence, p_1. Accordingly, the four positions adjacent to the left-hand face assign p_1 the value false, and the remaining four positions adjacent to the right-hand face consider p_1

true. Likewise, the y-dimension and the z-dimension correspond to the truth values of p_2 and p_3, respectively: p_2 (p_3) is true according to the positions adjacent to the front face (upper face) of the cube. We consider, in order to illustrate the effect of different argumentation strategies, two proponents. The proponent who puts forward a new argument (in line with one of the argumentation strategies) adopts the position $[p_1, p_2, p_3]$, that is, regards all sentences as true. In the cases we will examine, her opponent assumes various positions, which are, however, always distinct from the proponent's position.

Figure 6.5 provides, for each argumentation strategy, an example of how an argument which is put forward in line with the respective argumentation rule shapes the space of coherent positions. In Fig. 6.5a, the proponent *fortifies* her own position by introducing the argument (p_1,p_2;p_3). As a consequence, the position that considers p_1 and p_2 true, yet p_3 false, becomes incoherent. In panel (b), the proponent introduces, following the *convert* rule, the argument ($\neg p_2$;p_3); she starts from a premiss the opponent agrees with and derives a sentence she regards as true herself (agreeing, coincidentally, with the opponent). Relying but on one premiss, the new argument renders two positions incoherent. The argument (p_1;p_2) serves as an example for an *attack* on the opponent position, as shown in Fig. 6.5c. It takes off from the proponent's position and demonstrates that a controversial belief of the opponent (namely, that $\neg p_2$) is false. Finally, panel (d) depicts the effect of the argument ($\neg p_1$;p_2), which *undercuts* the opponent by explicating an internal inconsistency within her position. Every argument that undercuts a position renders that position incoherent. According to the *closest coherent* update mechanism, the opponent will readjust her position to either $[\neg p_1, p_2, p_3]$ or $[p_1, \neg p_2, p_3]$.

So as to generalize these examples, we identify, for the four cases just studied, *all* the positions which may be rendered incoherent by introducing some argument (with two distinct premisses) that satisfies the corresponding argumentation rule. So, Fig. 6.6a depicts those positions which the proponent may eliminate from the space of coherent positions by following the *fortify* rule. Likewise, panels (b)–(d) pinpoint the positions that can be rendered incoherent by the remaining three strategies. As a first thing to observe, the positions removable by an *attack* or an *undercut* may be deleted by *fortifying* or, respectively, *converting*, as well: This simply reflects the fact that every *attack* is eo ipso a *fortification* and *undercutting* implies *converting*. More interestingly, the strategies that take the proponent's beliefs as premisses, *fortify* and *attack* that is, eliminate positions in the vicinity of the proponent position itself, resulting in its isolation. Moreover, by *attacking* an opponent's position, a proponent seems to delete precisely those positions in her neighborhood which are particularly close to the opponent. Future rapprochement is thus effectively forestalled. The strategies which choose the opponent's position as point of departure, however, potentially eliminate positions in the vicinity of the opponent position, including the opponent position itself. Still, the *convert* strategy, as shown in panel (b), leaves the positions which are (1) contiguous to the opponent position and, in the same time, (2) comparatively close to the proponent position intact. Only neighboring positions which are even more distanced from the

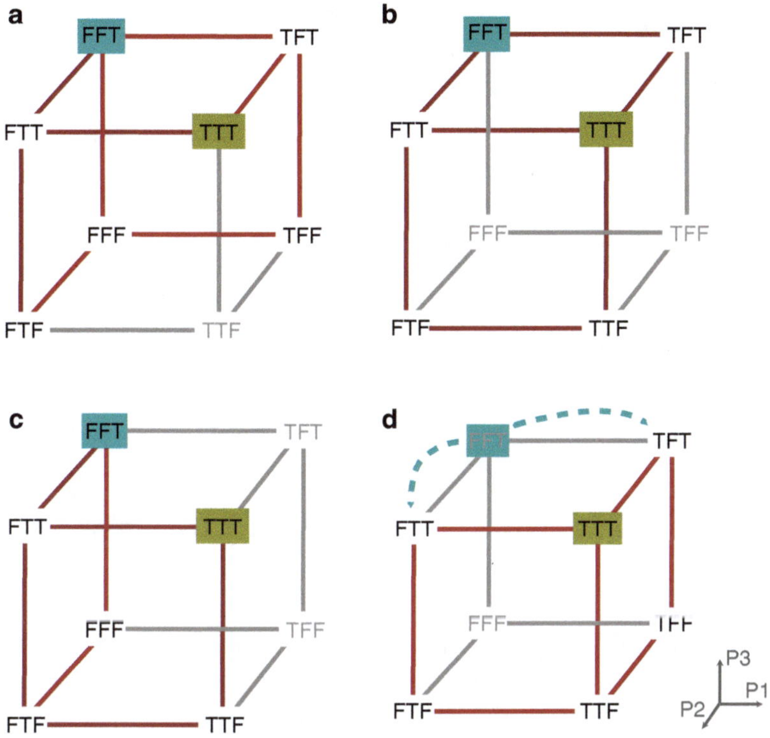

Fig. 6.5 Effects of introducing an argument in line with the four argumentation strategies on a small space of coherent positions. The sentence pool contains 2×3 sentences. All combinatorially possible positions are, initially, coherent. Positions are displayed as sequences of truth values, for example, "FFT" represents the position according to which p_1 and p_2 are false, and p_3 is true. In this 3-dimensional visualization, the *x*-, *y*-, and *z*-dimension correspond to exactly one of the three sentences being true or false. Proponent (khaki) and opponent (turquoise) assume different positions. Coherent positions which are rendered incoherent due to the newly introduced argument are tinted gray. Panel (**a**): The proponent introduces the argument $(p_1, p_2; p_3)$, *fortifying* her position. Panel (**b**): The proponent tries to *convert* her opponent by putting forward $(\neg p_2; p_3)$. Panel (**c**): The proponent *attacks* her opponent with $(p_1; p_2)$. Panel (**d**): The proponent advances $(\neg p_1; p_2)$, *undercutting* the opponent position

proponent than the opponent's current position can be eliminated by the *convert* strategy. These examples suggest that *convert* and *undercut* might indeed be much more effective strategies for reaching agreement than *fortify* and *attack*. In any case, if a proponent were able to put forward all arguments which correspond to the *convert* rule at once, the opponent would necessarily approach the proponent's position (given no arguments had been introduced before). This is not true of any other argumentation strategy.

In a next step, we generalize the examples from above even further. In order to do so, we have to develop a more abstract understanding of how arguments shape the

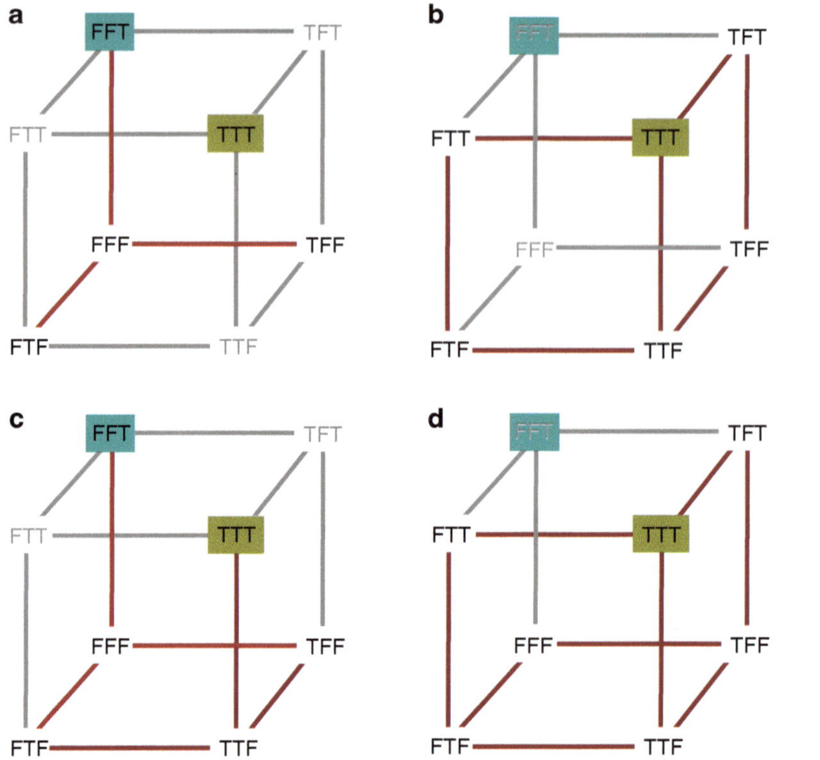

Fig. 6.6 Positions which the proponent (khaki) can render incoherent by putting forward arguments with two distinct premisses in line with the four argumentation strategies. See Fig. 6.5 for a more detailed explanation of this kind of visualization. Panels show the potential effects of different argumentation strategies. Panel (**a**): *fortify*. Panel (**b**): *convert*. Panel (**c**): *attack*. Panel (**d**): *undercut*

space of coherent positions quite generally. In brief, introducing an argument with k premisses into a debate (with a sentence pool of size $2n$) translates, in terms of the space of coherent positions, into:

1. Selecting a $(n-k)$-dimensional subspace of the entire space of coherent positions. This subspace contains all coherent positions according to which the k premisses are true.
2. Cutting the entire space of coherent positions (SCP), and thence also the selected subspace, into two parts: positions according to which the conclusion is true on the one side, and positions according to which the conclusion is false on the other side.
3. Removing all positions in the selected subspace (step 1) which also belong to that part of the SCP (step 2) where the conclusion is false. These are precisely the positions rendered incoherent by the new argument.

The four argumentation strategies can now be described neatly in terms of how the respective $(n-k)$-dimensional subspace is selected (step 1) and how the SCP is cut into two halves (step 2).

Fortify: Choose the $(n-k)$-dimensional subspace so that the *proponent position* lies within that subspace. Cut the space of coherent positions such that the *proponent position* lies within the half where the conclusion holds.

Convert: Choose the $(n-k)$-dimensional subspace so that the *opponent position* lies within that subspace. Cut the space of coherent positions such that the *proponent position* lies within the half where the conclusion holds.

Attack: Choose the $(n-k)$-dimensional subspace so that the *proponent position* lies within that subspace. Cut the space of coherent positions such that (1) proponent and opponent position reside in different halves and (2) the *opponent position* lies within the half where the conclusion is false.

Undercut: Choose the $(n-k)$-dimensional subspace so that *opponent position* lies within that subspace. Cut the space of coherent positions such that (1) proponent and opponent position reside in different halves and (2) the *opponent position* lies within the half where the conclusion is false.

Figure 6.7 illustrates these alternative descriptions of the four strategies with regard to our example of a 3-dimensional space of coherent positions. In panel (a), the proponent *fortifies* her position with $(p_1;p_2)$. Accordingly, all positions which extend $[p_1]$ belong to the $(3-1)$-dimensional subspace, including the proponent position itself. This subspace is cut into two halves (positions that consider p_2 true on the one side and false on the other side), and one half is declared as incoherent such that the proponent position lies within the coherent half. In panel (b), the proponent introduces $(\neg p_1;p_2)$ in line with the *convert* rule. She thus selects a subsection comprising the opponent position, namely, the set of all positions according to which p_1 is false, and cuts the entire SCP in two parts—in one part, p_2 holds, and in the other one, it doesn't. The location of the proponent position determines which of these two halves contains the positions that are rendered incoherent: The positions in the 2-dimensional subsection which remain coherent reside in the very same half of the SCP as the proponent position. In this particular example, the opponent position does so, too. It remains, as a consequence, coherent. As panel (c) illustrates, putting forward an argument in line with the *attack* rule is basically equivalent with *fortifying* one's position, except that, in order to count as an *attack*, a *fortify* argument has to cut the entire space of coherent positions (transparent surface) so that opponent and proponent assume positions in different halves. In this panel, the proponent introduces $(p_1;p_2)$. Because the opponent occupies, in contrast to case (a), the position $[\neg p_1, \neg p_2, p_3]$, this represents an *attack*. Finally, panel (d) shows that the *undercut* strategy determines the $(3-1)$-dimensional subspace and the cut of the SCP in roughly the same way as the *convert* rule. However, the *undercut* strategy requires, in addition, that the SCP be cut so that the proponent's and opponent's positions lie on opposing halves. With the opponent adopting, in panel (d), the position $[\neg p_1, \neg p_2, p_3]$, putting forward $(\neg p_1;p_2)$ amounts to undercutting the opponent and renders her position, unlike in case (b), incoherent.

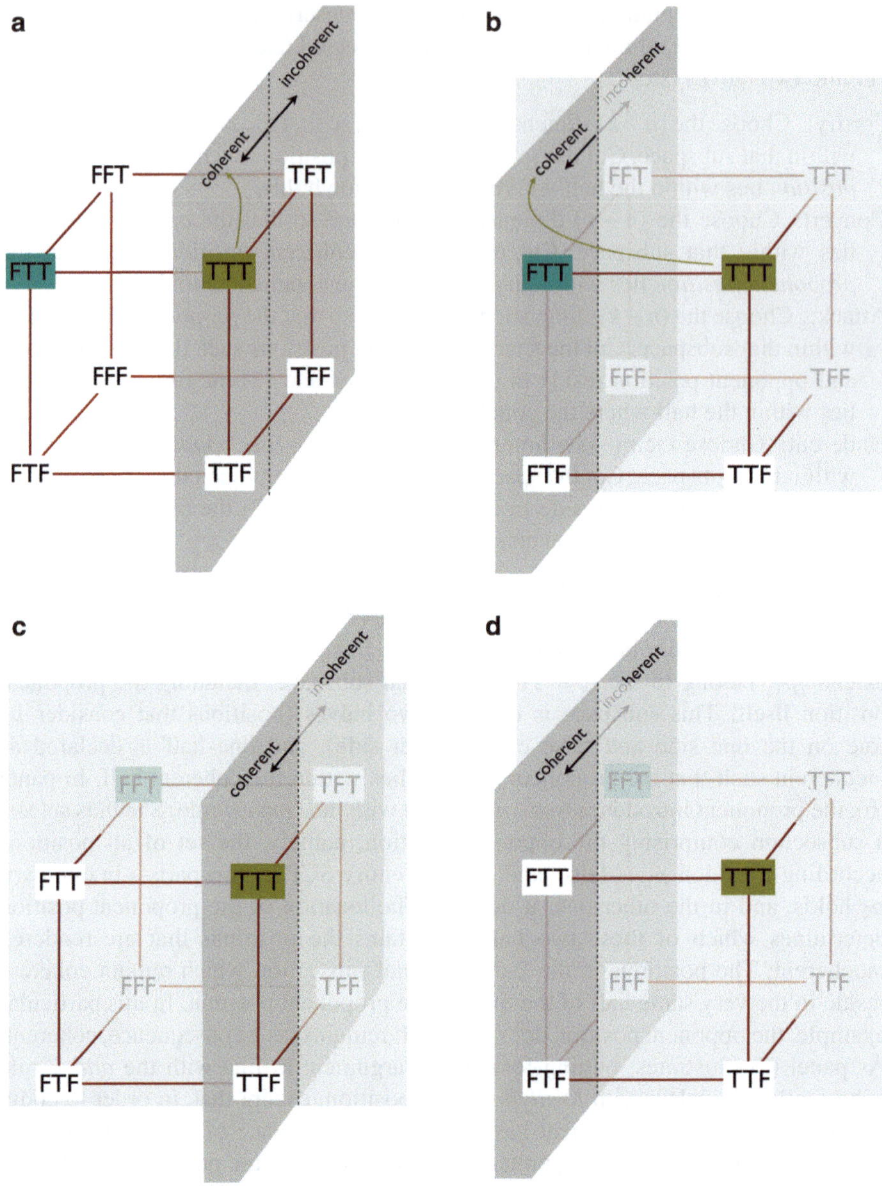

Fig. 6.7 Illustration of the four argumentation strategies in terms of the abstract description how the positions to be eliminated are determined. The space of coherent positions is visualized as in Fig. 6.5. The $(n-k)$-dimensional subspace which a proponent selects by choosing the premisses of her argument is represented by a $(3-1)$-dimensional subspace—a surface (*dark gray*). We visualize, accordingly, the effect of introducing an argument with one premiss. By specifying a conclusion, the entire space of coherent positions, including the gray surface, is cut in two halves. This cut is depicted by a transparent surface in *light gray* (in panels **b**–**d**), and the cut of the $(3-1)$-dimensional subspace is, in addition, indicated by a *dashed line*. The four panels illustrate how the four argumentation strategies differ in regard of determining the $(3-1)$-dimensional subspace and the cut of the SCP. Panel (**a**): *fortify*. Panel (**b**): *convert*. Panel (**c**): *attack*. Panel (**d**): *undercut*

Our abstract account of the argumentation rules underpins, in a general way, the specific observations we have previously made. In particular, we are now in a position to understand that the argumentation strategies which select convictions of the proponent as premisses, namely, *fortify* and *attack*, result in an isolation of the proponent position by rendering nearby positions incoherent. This is because the $(n-k)$-dimensional subspace from which some positions are eliminated is chosen such that the proponent position lies within that space. The opponent's position is only rendered incoherent by chance, namely, when it lies, coincidentally, in the respective section close to the proponent position. The argumentation strategies that prescribe to rely on premisses the opponent considers true, in contrast, lead to the removal of positions in the vicinity of the opponent's position, including, possibly, the opponent position itself. Because the *convert* and *undercut* rule stipulate to delete only coherent positions which don't agree, as regards the conclusion, with the proponent position, they prescribe to remove precisely that half of the $(n-k)$-dimensional subspace which is farther apart from the proponent position. The opponent position is thence systematically pushed towards the proponent.

Equipped with these conceptual insights, let us return to the stylized facts and try to explain them. To start with, the greater consensus-conduciveness of *convert/undercut* as opposed to *fortify/attack* (2.a, 4.a) can be made intelligible along the following lines: As we have seen, *fortify/attack* tend to isolate the proponent's position in the SCP, forestalling gradual rapprochement, whereas *convert/undercut* shape the SCP so that opponents are systematically pushed to the proponent's position. Moreover, only *convert/undercut* explicitly try to render opponent positions incoherent at all, forcing them to modify their position—no matter in which direction—in the very first place. No wonder *convert/undercut* turn out to be more effective with regard to increasing mean agreement and bringing about full consensus. The fact that *fortify/attack* tend to leave opponent positions intact allows us to explain why argumentation contexts that are made up of these strategies exhibit a relatively low versatility (5.a). In such contexts, proponents hardly alter their positions because the arguments, advanced by their opponents, don't force them to do so. And the opposite is true of argumentation contexts with *convert/undercut*. This brings us to the observation that an opponent's strategy exerts a major influence on the proponent's versatility (5.c). Consider a proponent whose opponent follows the *undercut* rule. This implies that with every new argument the opponent introduces, the proponent's position is rendered incoherent and has to be modified. If, however, the opponent applies the *fortify* strategy, she doesn't take into account the proponent's position at all, scarcely ever forcing the proponent to readjust her position.

Let us next turn to the more detailed stylized facts. The *fortify* rule outperforms the *attack* strategy with respect to mean agreement increase and consensus-conduciveness (2.b, 4.b). This difference can be understood by considering the specific way these rules shape the space of coherent positions. So, not only does, as noted above, the *attack* strategy result in an isolation of the proponent position by eliminating positions located nearby. What is more, arguments which conform with the *attack* rule eliminate precisely those coherent positions that represent intermediate positions in between the proponent and her opponent—that is, candidates for compromises. In addition, and with this we move on to the role of initial agreement, the *attack* rule turns out to be very effective in destroying high initial agreement (3.a). Unlike the *fortify* rule, it pushes positions apart. Obviously, this drives the poor general performance of the *attack* strategy (2.b, 4.b). But why does, for example, *attack–attack* cause a loss of agreement? The crucial difference to the *fortify* rule is that the conclusion of an argument which conforms with the *attack* rule is always denied by the opponent. As a consequence, opponent positions are more likely to be rendered incoherent if the proponent applies *attack* rather than *fortify*. But besides denying its conclusion, an opponent obviously has to assent to an argument's premisses, which are, in the case of *fortify*/*attack*, accepted by the proponent, so as to be compelled to modify her position. Therefore, closeness to a proponent who implements *attack* increases the likelihood of being rendered incoherent, too. This explains, together with the fact that the *attack* rule isolates positions and thence prevents opponent positions, once falsified, from moving into the direction of the proponent, why the *attack* strategy drives proponents apart. It follows, besides, that neither *attack* nor *fortify* can render very distant positions incoherent, since those don't share the premisses the proponent adheres to in the first place. As a result, drastic disagreement isn't resolved by those rules (3.b).

Despite their general resemblance, *convert* and *undercut* exhibit some notable differences, as well. Our conceptual investigations above have already suggested why *convert* is more effective as a rule for increasing mean agreement than *undercut* (2.c): It is the only rule which, provided all arguments that conform with the rule are put forward at once, forces the opponent to approach the proponent. The fact that the *convert* strategy does not imply, unlike the *undercut* strategy, that every new argument renders the opponent position incoherent helps to explain the success of *convert*, too. As the *undercut* strategy forces the opponent to readjust her position with every new argument, it is barely able to control or to influence the way the opponent modifies her position. The *convert* strategy, however, allows the proponent to build up an argumentation which eliminates, in a first step, coherent positions the opponent might have considered as attractive (i.e., close) fallback positions, before, in a second step, directly targeting the opponent and now forcing her, with the fallbacks rendered incoherent, to move towards the proponent. This sophisticated strategy is, clearly, not built into the simple *convert* rule; yet, the *convert* rule at least allows for this kind of argumentation process, whereas the *undercut* rule doesn't. Because *undercut* addresses opponents much more aggressively, it risks to divide opponents—not having closed backdoors in the SCP—much more frequently than *convert*. This very same mechanism explains why *undercut* tends to undo coincidentally high initial agreement, whereas *convert* doesn't (3.c). Moreover, the frequent falsifications of opponent positions by the *undercut* rule foster, especially at low densities, the random walk effect we had already observed in random debates (cf. Sect. 4.3). This contributes both to the notable drop in mean agreement between positions with high initial agreement as well as to the steep agreement increase for initially very distant positions (3.c). But in spite of all these apparent advantages of the *convert* strategy, the *undercut* strategy seems to bring about full consensus more rapidly than the *convert* strategy (4.c). How does this fact fit into the picture we've drawn so far? I suggest that *convert* is a very effective rule for bringing positions closer together; it might, however, not be the most effective strategy for triggering the final step from high agreement to full consensus. This is because if two positions agree largely, yet not fully, applying the *convert* strategy will, in most cases, result in an argument which simply *fortifies* the opponent position: As the conclusion of such an argument is a randomly chosen conviction of the proponent, the opponent is very likely to agree with it, too. Such arguments don't cause the opponent to modify her position and therefore don't resolve the residual disagreement. Arguments which conform with the *undercut* strategy, in contrast, directly address the remaining agreement and are thus rather in a position, at least in the later phase of a debate, to remove remaining dissent rapidly.

Let us, finally, consider how the different rules compare in terms of versatility. Note that we can estimate the average likelihood that an argument, conforming with one of the argumentation strategies, renders the opponent position incoherent: With *undercut*, the opponent position is always rendered incoherent. With *convert*, the opponent position is rendered incoherent if and only if the opponent coincidentally disagrees with the argument's conclusion. With *attack*, the opponent position is rendered incoherent if and only if the opponent coincidentally agrees with all (two) premisses. And with *fortify*, finally, the opponent position is rendered incoherent if and only if the opponent agrees, coincidentally, with all (two) premisses while disagreeing, in the same time, with the conclusion. In general, the likelihood that a proponent assigns, coincidentally, certain truth values to k different sentences decreases with k. Still, if a proponent renders her opponent's positions incoherent more often than the opponent falsifies, vice versa, the proponent's positions, then the opponent has to modify her position more frequently and will, as a result of these modifications, eventually depart from her original position to a larger extent. Collectively, these observations explain why *undercut* dominates *convert*, *convert* dominates *attack*, and *attack* dominates *fortify* (5.b).

Chapter 7
The Consensual Dynamics of Argumentation Strategies in Many-Proponent Debates

In the previous chapter, we have carried out a comparative analysis of different argumentation strategies by simulating dualistic (two-proponent) debates. As noted above, the results of these simulations may not be directly scaled up to debates with more than two proponents. This is the reason why we study, in this chapter, debates with six proponents who implement argumentation strategies which are derived from the basic rules previously introduced.

7.1 Setup

More specifically, we consider two different argumentation strategies: *multiple convert* and *multiple undercut*. Both are slightly modified versions of the two most consensus-conducive strategies studied in the previous chapter. We set up two ensembles (with 1,000 debates)—one for each argumentation strategy. There are six proponents per debate. The simulations terminate as soon as all proponents come to agree, or the inferential density rises above 0.8.

The precise specification of debates in the *multiple undercut* ensemble is:

Argumentation Mechanism: Proponents, in alternating sequence, introduce arguments into the debate according to a modified *undercut* strategy. A proponent i, when it's her turn, first of all identifies a sentence c she considers true and most of her opponents don't agree with. In other words, she selects a sentence $c \in S$ such that (1) $\mathscr{P}_t^i(c) = \text{true}$ and (2) $|\{j|\mathscr{P}_t^j(c) = \text{false}\}|$ is maximal. In a second step, she determines all pairs of sentences (excluding $c/\neg c$), such that the number of opponents who accept both sentences yet disagree with c is maximal. Formally, these two distinct sentences $p_1, p_2 \in S \setminus \{c, \neg c\}$ maximize $|\{j|\mathscr{P}_t^j(c) = \text{false} \wedge \mathscr{P}_t^j(p_1) = \mathscr{P}_t^j(p_2) = \text{true}\}|$. The proponent now introduces an argument with conclusion c and one of these pairs of sentences as premisses—taking into

G. Betz, *Debate Dynamics: How Controversy Improves Our Beliefs*, Synthese Library 357, DOI 10.1007/978-94-007-4599-5_7,

account the extra condition that adding this argument to τ_t leaves at least one position coherent. This gives rise to τ_{t+1}. We shall refer to this argumentation strategy as *multiple undercut*.

Discovery Mechanism: The background knowledge $\mathscr{B}$ is empty.

Update Mechanism: *Closest coherent* (cf. Sect. 4.1).

Debates in the *multiple convert* ensemble are set up as follows:

Argumentation Mechanism: Proponents, in alternating sequence, introduce arguments into the debate according to a modified *convert* strategy. A proponent i chooses randomly, in a first step, a sentence c she considers true ($\mathscr{P}_t^i(c) = \text{true}$). In a second step, she determines all pairs of sentences (excluding $c/\neg c$) such that the number of opponents who accept both sentences is maximal. Technically, the two distinct sentences $p_1, p_2 \in S \setminus \{c, \neg c\}$ maximize $|\{j | \mathscr{P}_t^j(p_1) = \mathscr{P}_t^j(p_2) = \text{true}\}|$. The proponent now introduces an argument with conclusion c and one of the sentence pairs as premisses—taking into account the extra condition that adding this argument to τ_t leaves at least one position coherent. This gives rise to τ_{t+1}. We shall refer to this argumentation strategy as *multiple convert*.

Discovery Mechanism: The background knowledge $\mathscr{B}$ is empty.

Update Mechanism: *Closest coherent* (cf. Sect. 4.1).

7.2 Results

Figure 7.1 compares, for our two ensembles, the mean agreement evolutions (1) averaged over all position pairs and (2) averaged over position pairs with specific initial agreement. In both ensembles, total mean agreement (left-hand panels) drops initially. In the *multiple convert* ensemble, this decline amounts to 2%. The trend is, however, quickly reversed even before a density of 0.1 is reached, after which mean agreement rises at a constant rate. At a density of 0.5, it has increased by more than 20 percentage points. This represents a significantly stronger rapprochement than in debates with *random argumentation* where agreement has increased, at a density of 0.5, by merely 5% (cf. Sect. 4.2). In the *multiple undercut* ensemble, in contrast, mean agreement falls, initially, by 5% to 0.45. This initial drop is recouped only slowly. Thus, proponents disagree with respect to more than half of the sentences unless an inferential density of 0.35 is attained. Consequently, mean agreement is even substantially lower, during this first phase, than in random debates. At a density of 0.5, mean agreement amounts to barely 0.6—considerably less than in the *multiple convert* ensemble. No earlier than at a density of 0.8, when proponents agree with regard to 90% of the sentences, *multiple undercut* catches up with *multiple convert*.

The overall trend is nicely reflected in the agreement evolutions regarding position pairs with specific initial agreement, too (right-hand panels in Fig. 7.1). In both ensembles, the mean agreement of positions with medium initial agreement (middle curves) follows closely the corresponding total mean agreement evolution.

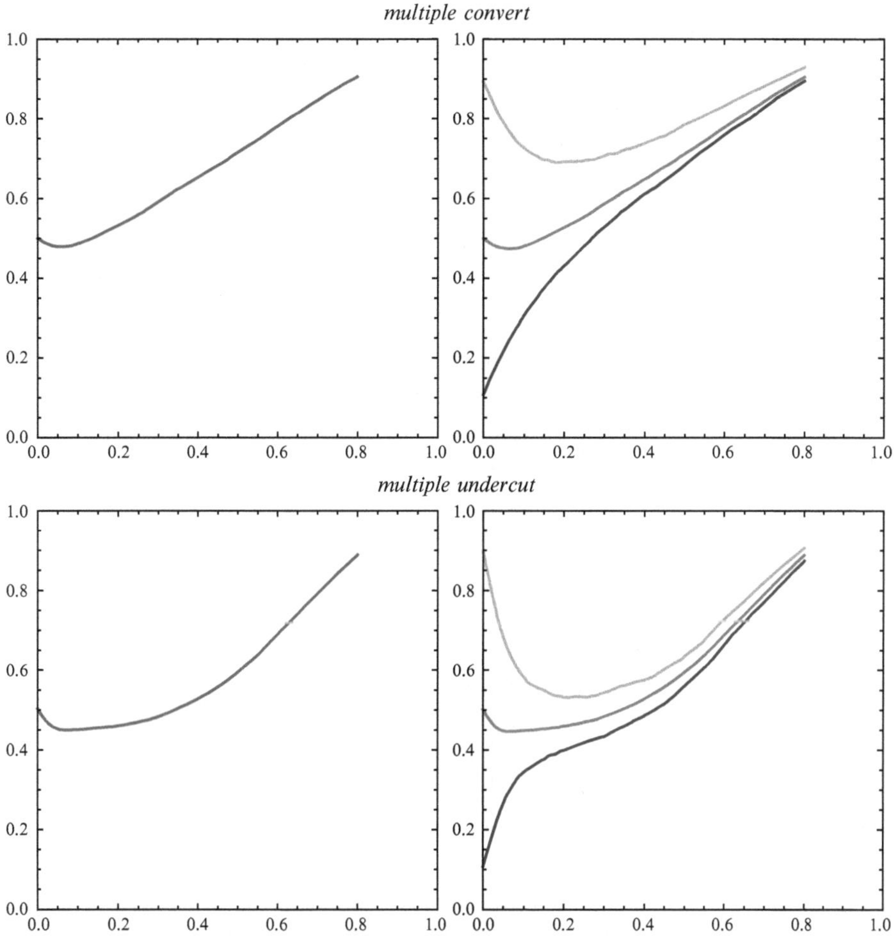

Fig. 7.1 Ensemble-wide mean agreement as a function of inferential density, averaged over all position pairs (*left*) and position pairs with specific initial agreement (*right*)

Positions with extreme initial agreement display, in both ensembles, a substantial initial increase (high initial disagreement, bottom curves) or decline (high initial agreement, top curves). The early change in mean agreement is, however, much stronger in the *multiple undercut* than in the *multiple convert* ensemble. So, in debates with *multiple undercut*, high initial agreement evaporates much more rapidly and thoroughly—falling by 0.35. Likewise, positions which are initially far apart approach each other, at low densities, faster, if proponents apply *multiple undercut* instead of *multiple convert*. Yet, the early elimination of severe disagreement (bottom curve) doesn't suffice to compensate the loss of coincidentally high agreement in the *multiple undercut* ensemble.

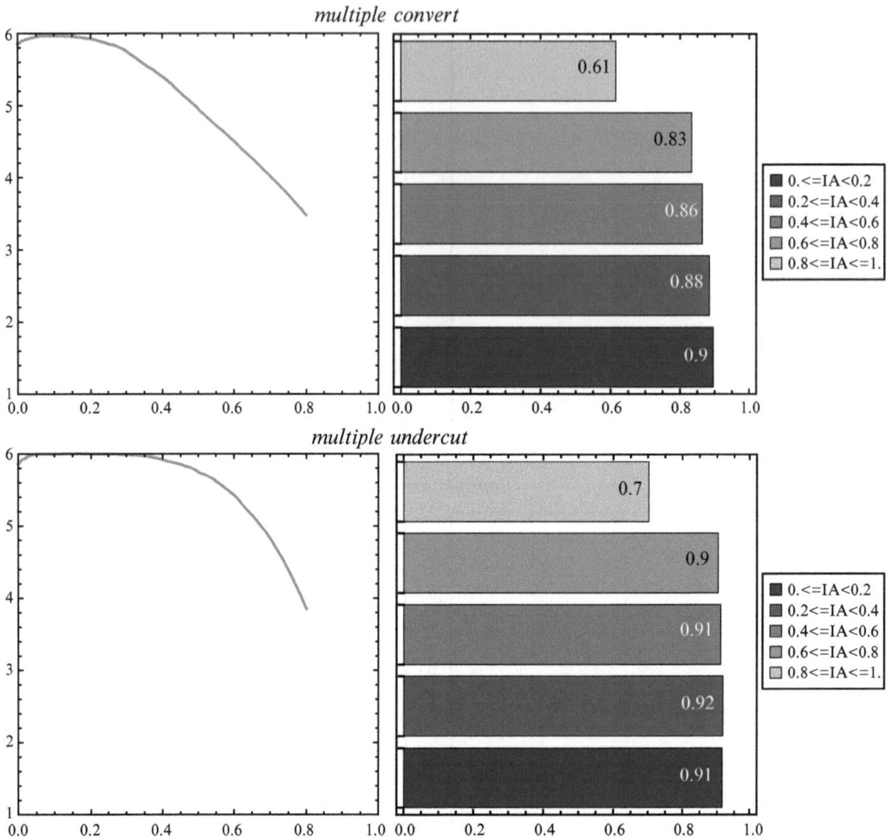

Fig. 7.2 Ensemble-wide average number of nonidentical positions as a function of inferential density (*left*) and mean collapse densities for position pairs with specific initial agreement (*right*). To calculate the average collapse densities, we assume that position pairs which haven't come to full agreement when the simulation terminates ($D > 0.8$) reach consensus at $D = 1$

Figure 7.2 provides an alternative perspective on the position dynamics in our two ensembles—telling us how fast proponents reach complete consensus in the debates. As the left-hand panels demonstrate, the number of nonidentical positions starts to fall earlier in debates with *multiple convert* and, as a result, lies constantly below the corresponding value in the *multiple undercut* ensemble. So, in contrast to the findings in the previous chapter, *multiple convert* is not only more effective in terms of increasing mean agreement but also with respect to generating full consensus. A comparison of the mean collapse densities (right-hand panels) corroborates this result. No matter what the initial agreement of two positions, they tend to reach full consensus at lower densities in the *multiple convert* as compared to the *multiple undercut* ensemble.

Table 7.1 Fragmentation of the SCP, measured by aggregated NCC, in different ensembles

Ensemble	Lower 10th quantile	Ensemble-wide mean	Upper 10th quantile
random argumentation	1.047	1.087	1.132
multiple undercut	1.028	1.058	1.099
multiple convert	0.931	1.011	1.067

During the discussion of position dynamics of debates with *random argumentation* (see Chaps. 4 and 5), the degree of fragmentation of the space of coherent positions turned out to be a relevant factor for understanding the different evolutions of mean agreement and number of nonidentical positions. So, how fragmented are the debates in this chapter's two ensembles, and how do debates with different degrees of fragmentation evolve? We have quantified the fragmentation of the SCP by the aggregated NCC of the six proponent positions in the density interval $[0;0.5]$. Debates with high aggregated NCC possess a relatively compact and well-connected SCP, and debates with low aggregated NCC, in contrast, display high fragmentation. Now, comparing the random debates studied in Chap. 4 with the *multiple convert* and the *multiple undercut* ensemble, the aggregated NCC, averaged over all debates in the corresponding ensemble, is much lower in this chapter's ensembles (see Table 7.1). In other words, *multiple undercut* and *multiple convert* give rise to much more fragmented debates, or, more precisely, the SCP is more fragmented *and* proponents occupy more remote positions. This holds in particular for *multiple convert*: The average aggregated NCC in the *multiple convert* ensemble lies not only below the average in the *multiple undercut* ensemble but lies even substantially below the latter's lower 10th quantile. In addition, aggregated NCC of the most fragmented debates (lower 10th quantile) in the *multiple convert* ensemble differs from the ensemble's mean by 0.08 points, as compared to corresponding differences of 0.03 and 0.04 in the other ensembles. In sum, *multiple convert* appears to be highly effective with regard to fragmenting the SCP and causing proponents to hold remote positions.

How do these differences in terms of absolute degree of fragmentation translate into features of the position dynamics? Figure 7.3 provides an answer to this question by displaying the evolutions of mean agreement and number of nonidentical positions averaged over all debates (solid curves), highly fragmented debates (dashed curves), and very compact debates (dotted curves). Let us consider the *multiple undercut* ensemble, first. Here, things look pretty familiar. As in the case of *random argumentation* (compare Figs. 4.6 and 4.7), compact debates display a more stable and slightly earlier mean agreement increase (dotted curve, left-hand panel), whereas fragmented debates are characterized by below-average mean agreement (dashed curve). In fragmented debates, however, proponents tend to reach consensus at lower densities than in compact ones (the dashed curve in the right-hand panel lies below the dotted one). So, like in the *random argumentation* ensemble, fragmentation fosters full consensus while being relatively unfavorable to gradual rapprochement. In the *multiple convert* ensemble, which we shall consider

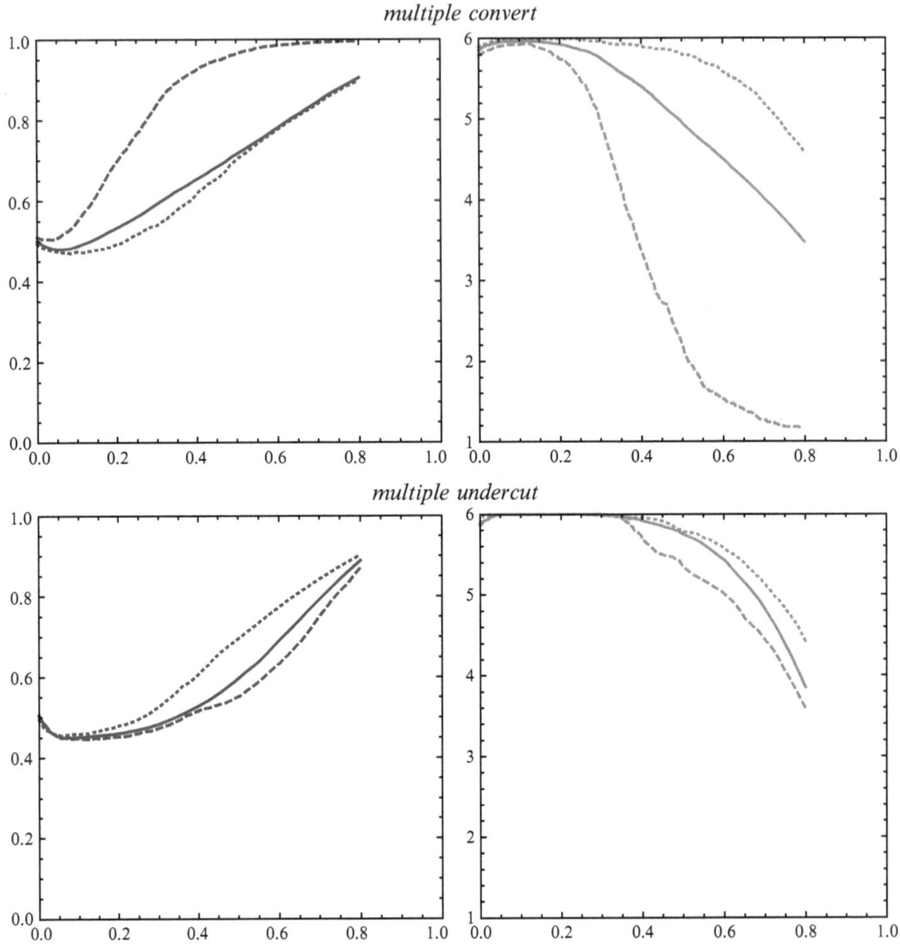

Fig. 7.3 Ensemble-wide mean agreement evolutions (*left*) and average number of nonidentical positions (*right*) in fragmented and compact debates as functions of inferential density. The different curves are calculated by taking into account all debates (*solid*), debates with aggregated NCC smaller than lower 10th quantile (*dashed*) and debates with aggregated NCC greater than upper 10th quantile (*dotted*)

next, the most fragmented debates behave completely differently. As the two plots at the top Fig. 7.3 demonstrate, mean agreement is quickly propelled to levels way above the ensemble mean; likewise, the number of nonidentical positions plunges rapidly to very low values. More specifically, at a density of 0.5, proponents agree, on average, with respect to more than 95% of the sentences in a fragmented debate. At the same density, the six proponents occupy, on average, merely two different positions. Whereas the position dynamics of fragmented debates deviates

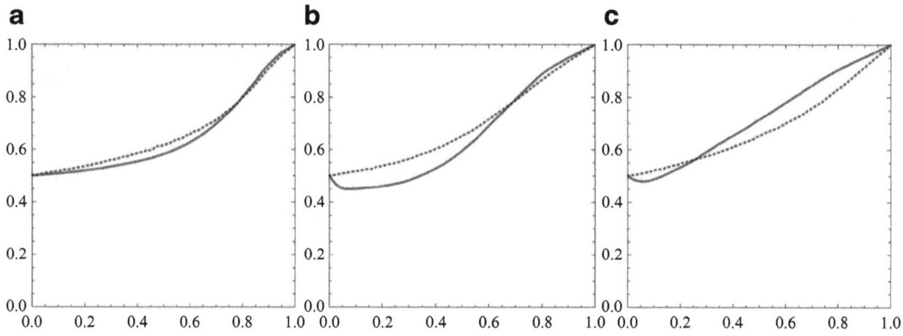

Fig. 7.4 Mean agreement evolution of consecutively updated, simulated proponent positions (*solid curves*) compared with mean agreement of random positions that are newly chosen at each step of the debate (*dotted curves*), plotted against inferential density. Panel (**a**): Random debate simulations as presented in Chap. 4. Panel (**b**): Simulations of debates with *multiple undercut* strategy. Panel (**c**): Simulations of debates with *multiple convert* strategy. Mean agreement of random positions at some inferential density *D* (*dotted curves*) is calculated by averaging the mean agreement of 24 randomly chosen coherent positions at the density *D* over all debates of the respective ensemble. Mean agreement values above a density of 0.8 are extrapolated linearly in panels (**b**) and (**c**)

substantially from the corresponding evolutions in the *multiple undercut* ensemble, the compact debates (dotted curves) exhibit in both ensembles—qualitatively as well as quantitatively—very similar behavior. Therefore, the observed differences between ensemble-wide mean agreement evolutions seem to result, primarily, from the different evolutions of highly fragmented debates.

The final finding reported in this section probes the general cause of rapprochement in the ensembles studied so far. The increase in mean agreement we have observed in this chapter's ensemble as well as in other ones can stem from two different effects. On the one hand, it might simply be due to the contraction of the SCP. If the SCP shrinks, so does the distance between two arbitrarily chosen coherent positions. And this causes proponent positions to approach each other, as well. Differences in speed of rapprochement might, accordingly, be explained in terms of how effectively argumentation mechanisms reduce the mean distance between coherent positions in the SCP. On the other hand, proponent agreement can be generated by pushing proponents together and trying to gather them in certain parts of the SCP. This might be achieved, in principle, without reducing the mean distance between two arbitrary positions in the SCP at all. The question is: How strong are these two different effects? Figure 7.4 compares, for *random argumentation*, *multiple convert*, and *multiple undercut*, the ensemble-wide mean agreement (solid curves) of proponent positions on the one side with the ensemble-wide mean agreement of randomly chosen coherent positions at the corresponding density (dotted curves) on the other side. The latter curves thence represent the (hypothetical) mean agreement of newcomers who adopt randomly chosen coherent positions at the respective state of the debate.[1] Interestingly, both with the *random argumentation* (panel a) as well as with the *multiple undercut* strategy (panel b), the six proponents in the debate agree significantly less than six randomly chosen newcomers would. Only at very high densities, the proponents agree to a slightly greater extent than randomly chosen coherent positions. The opposite is true for the *multiple convert* strategy. Whereas the mean agreement lies, initially, below the agreement of randomly picked coherent positions, the incumbent proponents reach substantially greater agreement than random newcomers beyond a density of 0.25.

7.3 Discussion

The results presented in the previous section can be summarized in the following, more general observations, which are to be explained henceforth:

1. The *multiple undercut* and the *multiple convert* strategy are, all in all, more consensus-conducive than purely *random argumentation*. This holds in any case for densities above 0.5 and, as regards *multiple convert*, for lower densities, as well.
2. High initial agreement is quickly reduced, while proponent positions with extreme disagreement approach each other rapidly. These early changes are more pronounced in the *multiple undercut* ensemble.
3. Mean agreement, averaged over all position pairs, is initially destroyed in both ensembles. This initial drop is particularly plain, and only slowly recouped, in the *multiple undercut* ensemble.
4. In the *multiple convert* ensemble, fragmented debates give rise to a surprisingly high rapprochement of proponent positions.
5. Only in the *multiple convert* ensemble, mean proponent agreement exceeds the agreement of randomly chosen positions.

It is relatively easy to make sense of the first two observations listed above. *Multiple convert* and *multiple undercut* are modeled after the two most consensus-conducive strategies studied in the previous chapter. No wonder they outperform purely *random argumentation*. So, it's not their overall higher effectiveness in generating agreement which calls for an explanation but—specifically in the case of *multiple undercut*—the fact that, as detailed in the third observation, they fail to do so under certain boundary conditions. Concerning the second stylized fact, the initial evaporation of extremely high and extremely low initial agreement seems to be a result of the random walk effect we had previously observed. As long as the inferential constraints are rather loose, proponent positions follow a random walk whose speed is determined by the frequency at which proponent

[1]Note that the expected agreement of n randomly and independently chosen coherent positions ($n \geq 2$) does not hinge on n.

positions are rendered incoherent. This also explains the bigger initial change in the *multiple undercut* ensemble. For whenever an opponent position is targeted by an argument which satisfies the *multiple undercut* rule, it is rendered incoherent. *Multiple undercut* compels proponents to modify their positions more frequently than *multiple convert*, thus giving to rise to a stronger random walk effect.

Let us consider the third observation, next. Surprisingly, mean agreement drops, initially, by at least 5% with *multiple undercut*. How can this be explained? *Multiple undercut* urges proponents to introduce arguments so as to invalidate as many opponent positions as possible. However, two opponent positions can only be rendered incoherent by one and the same argument if they agree with regard to at least three sentences (namely, two premisses and one conclusion). Hence, as a first thing to note, this argumentation strategy is biased to target proponents with higher agreement. In addition, introducing such an argument forces the opponents addressed to modify—independently of each other—at least one of the truth values they assign to the argument's sentences, that is, sentences on which they previously agreed. Now, it is more likely that the opponents change different sentences rather than one and the same one. So, arguing in line with *multiple undercut* systematically destroys partial agreement among proponents. *Multiple convert* does so, too, but to a lesser extent, since not every argument that satisfies the requirements of the *multiple convert* rule renders the opponent's position incoherent. This is why we observe a significant drop of mean agreement below 0.5 in the *multiple undercut* ensemble, which is only slowly recovered.

As briefly noted in the previous section, the superior consensus-conduciveness of *multiple convert* seems to stem, mainly, from its ability to generate substantial rapprochement in very fragmented debates. But how come that (a) *multiple convert* gives rise to debates with a very high fragmentation value (aggregated NCC) and (b) that these debates are characterized by rapid agreement increase? I suggest that this can be explained by reference to the same mechanism which allowed us to understand the superior performance of the *convert* rule, studied in the previous chapter. Thus, like its simpler sibling, the *multiple convert* rule allows for closing backdoors in the SCP before opponent positions are invalidated. Less metaphorically, an argumentation which agrees with the *multiple convert* strategy may, firstly, eliminate all coherent positions in the opponents' vicinity which are relatively distant from the proponent's point of view and may, secondly, render the opponents' positions incoherent so as to compel them to approach the proponent. By mutually applying this strategy, eliminating each other's fallback positions, the proponents eventually push each other together, creating a comparatively isolated, ever shrinking cluster within the SCP. Let me repeat: The development just outlined is not built into the *multiple convert* rule. By coincidence, proponents who apply *multiple convert* could also consistently invalidate each other's positions, thus effectively arguing in line with *multiple undercut*. In such a debate, the mechanism previously sketched doesn't operate, and we wouldn't expect to observe above-average rapprochement. Yet, all this dovetails with our simulation results, as only some debates with *multiple convert*—namely, the very fragmented ones—exhibit outstanding agreement increase. So, my suggestion is that *multiple convert* allows

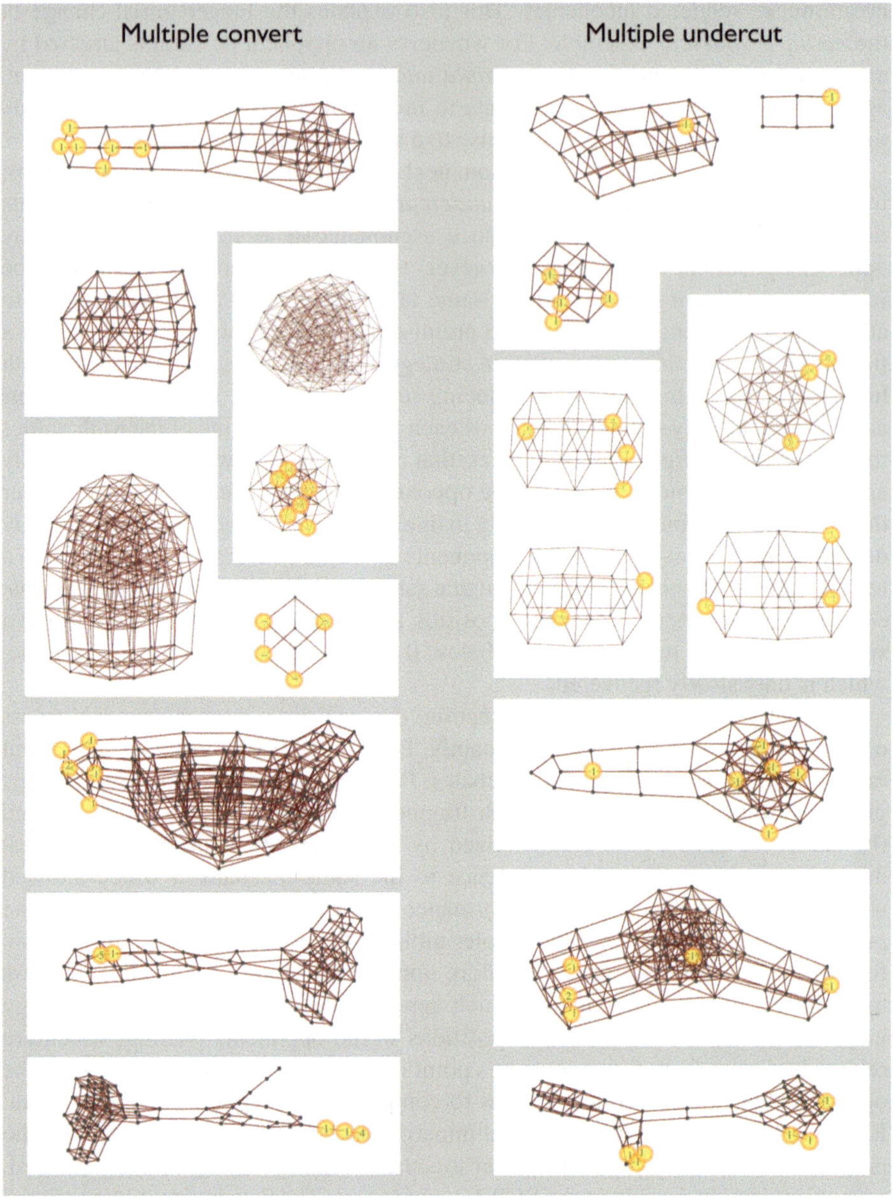

Fig. 7.5 Space of coherent positions of highly fragmented debates in the ensemble with *multiple convert* strategy (*left*) and *multiple undercut* strategy (*right*). The individual plots depict a 9-dimensional section of the SCP at step 45 (*left*) and at steps with somewhat higher corresponding densities (*right*). The plots represent an illustrative sample of the 100 most fragmented debates (low aggregated NCC) of each ensemble. Positions which are occupied by at least one proponent are highlighted by *yellow circles*

to gather all the proponent positions in gradually contracting subparts of the SCP, namely, in cases where proponents coincidentally, yet over and over again, eliminate fallback positions first before rendering opponent positions incoherent. In such a debate, proponents *construct* an opinion island which is populated by all proponents. Consequently, all proponents hold relatively remote positions, giving rise to a high aggregated NCC. Figure 7.5 corroborates this explanation. It displays, for each of our two ensembles, sections of six highly fragmented SCPs. In the examples drawn from the *multiple convert* ensemble, the proponent positions are tightly packed on a comparatively remote part of the SCP, whereas in the *multiple undercut* examples, proponent positions seem to be scattered all over the SCP. The 100 most fragmented debates in the *multiple convert* ensemble give, consistently, rise to very similar snapshots as shown in this figure. So, at least in some situations, *multiple convert* is very effective in generating agreement. These situations correspond exactly to the debates with a high aggregated NCC.

This last explanation sheds also some light on the fifth observation stated above. Whenever *multiple convert* gathers, as it sometimes does, the proponent positions in small subparts of the SCP, the mean agreement among proponents is significantly higher than the agreement of positions which are randomly chosen from the entire SCP. But remains the puzzle why, in the *multiple undercut* and *random argumentation* ensemble, proponents agree even less than randomly chosen positions. The fishing-net metaphor, introduced in Chap. 4, might, however, give us some clue: As the net is pulled—assuming that fishes are sluggish and were initially randomly distributed—more fishes will be located at the outer layers of the volume enclosed by the net (namely, all those fishes that have been forced to move) than at its center. Consequently, the density of fish exceeds, at the outsides, the density which would prevail in case of a completely random and homogeneous distribution. In the former case, fishes are thus, on average, farther apart than in the latter. The same holds, mutatis mutandis, for the flooded-village metaphor, as well. If we assume that people were initially randomly distributed all over the village and relocate only if forced to do so by the rising water, some villagers, namely those which have been displaced, will tend to gather where hills and roofs border on the water. As a result, two persons will be, on average, farther apart than two randomly chosen, nonflooded positions. By analogy, it is the inertia of the proponent dynamics—stemming from the fact that proponents move (1) slowly and (2) only as far as they have to in order to hold coherent positions—which causes mean agreement of proponent positions to lag behind average agreement of randomly chosen positions.

Chapter 8
The Consensual Dynamics of Debates with Core Updating

In the debates we have simulated so far, all sentences were on a par with each other. Proponents didn't consider some of the debates' sentences as more and others as less important. If they could, for example, reestablish coherency by changing exactly one truth-value assignment, they were indifferent as to which belief they give up. But this, it seems, doesn't hold in real controversies, where proponents frequently possess some convictions which they are very reluctant to give up as well as other beliefs they are much more willing to modify. In this chapter, we are going to include the proponents' varying loyalty to different beliefs in our simulations. In order to do so, we assume that there is a subset of the sentence pool which contains the debate's core sentences. The proponents' partial positions on these core sentences make up the heart of their belief system, which they are particularly unwilling to modify. In our simulations, this translates into a more sophisticated update mechanism, as explained below. Studying the simulation results, our primary concern is the evolution of the proponents' core positions. As these cores represent partial positions defined on a subset of the sentence pool, we will be able to examine how the robustness of positions, that is, the degree of justification, influences the debate dynamics.

8.1 Setup

We stipulate that five independent sentences of the sentence pool (plus their negations) form the key theses of a debate with respect to which proponents adopt their core positions. Core positions are thence partial positions of the proponents, defined on one and the same subset of the sentence pool. Proponents try to regain a coherent position, once their previous position has been rendered incoherent, without modifying their core beliefs.

We have 6 proponents per debate and a pool of $2 \cdot 20$ sentences. The specific debate setup is given below:

G. Betz, *Debate Dynamics: How Controversy Improves Our Beliefs*,
Synthese Library 357, DOI 10.1007/978-94-007-4599-5_8,

Argumentation Mechanism: *Random argumentation* (cf. Sect. 4.1).
Discovery Mechanism: The background knowledge remains empty.
Update Mechanism: Once τ_{t+1} is specified, it is checked (for every $i = 1 \ldots 6$) whether the proponent's complete position $\mathcal{Q}_t^i$ is coherent on τ_{t+1}. If it is, the complete position i remains unchanged ($\mathcal{Q}_{t+1}^i = \mathcal{Q}_t^i$). If it isn't, the proponent

1. Determines the coherent partial position $\mathcal{P}_{t+1}^i$, defined on the debate's core sentences, which is closest to her previous core position $\mathcal{P}_t^i$.
2. Finds the coherent extension of $\mathcal{P}_{t+1}^i$ to a complete position $\mathcal{Q}_{t+1}^i$ which is closest to her previous complete position $\mathcal{Q}_t^i$. This is her new proponent position.

In case there are several closest τ_{t+1}-coherent positions, in any of these steps, one of those is chosen randomly. We shall refer to this update mechanism as *lexicographic closest coherent*.

A debate simulation terminates if the six proponents hold identical core positions. We generate an ensemble of 1,000 debate simulations in line with the above specifications.

8.2 Results

Let us, first of all, verify that the modified update mechanism causes proponents to stick to their core beliefs more fiercely as compared to the simple *closest coherent* update. Figure 8.1 demonstrates that this is the case. It plots the distance between a proponent's core position at some density D and that proponent's initial core position—averaged over the entire ensemble. It visualizes, in other words, to which extent proponents are compelled to recede from their initial position as the inferential density increases. Obviously, proponents who update their positions in line with *lexicographic closest coherent* (bottom curve) hold on to their core beliefs more effectively than proponents who follow the simple *closest coherent* rule: Whereas, in both ensembles, a proponent's final core position at $D = 1$ disagrees, on average, with 50% of the proponent's initial beliefs, proponents with *lexicographic closest coherent* give in much more slowly—at a density of 0.5, for example, the core position of a proponent with *lexicographic update* disagrees by less than 10% with her original position, as compared to more than 30% with simple *closest coherent*.

The relatively high stability and inertia of core positions with *lexicographic closest coherent* are likely to translate into a slow rapprochement of the proponents' cores. Figure 8.2 confirms this expectation. Whereas in debates with simple *closest coherent* update (top curve), mean agreement among core positions evolves roughly at the same pace as total agreement (cf. Fig. 4.1), reaching, at $D = 0.5$, a level between 55% and 60%, core positions approach each other much more slowly

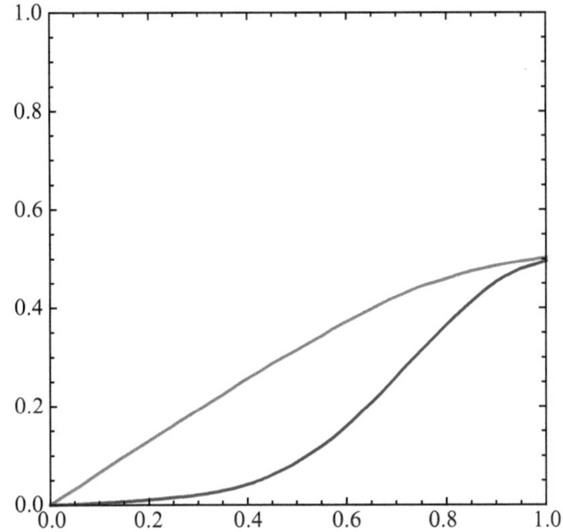

Fig. 8.1 Ensemble-wide mean normalized distance between current and initial core proponent positions as a function of inferential density. Mean distance to initial positions is plotted for this chapter's ensemble with *lexicographic closest coherent* update (*bottom curve*) and for the ensemble, presented in Chap. 4, with simple *closest coherent* (*top curve*). As regards the second case, the proponents' core beliefs are presumed to relate to the very same five sentences which make up the proponents' cores in the first ensemble. Note, however, that these core beliefs don't influence the debate dynamics if proponents update according to simple *closest coherent* and if arguments are introduced randomly

with *lexicographic closest coherent* update. At densities below 0.5, for instance, the agreement between the proponents' core beliefs hardly changes at all.

The difference between *simple* and *lexicographic closest coherent* is, however, less pronounced with respect to the number of nonidentical core positions, as Fig. 8.3 shows. In both ensembles, there are, due to coincidental agreement, initially 5.2 different core positions per debate, on average. In the ensemble with simple *closest coherent* update (b), the coincidental consensus is partially destroyed during the initial phase of the debates, before it starts to drop markedly at densities above 0.5. With *lexicographic closest coherent* (a), however, initial consensus does not evaporate (because of the cores' inertia), yet the eventual decline lags slightly behind the gray curve—full consensus is reached somewhat later with *lexicographic closest coherent*.

In the introductory paragraph of this chapter, we announced that we are going to use the concept of degree of justification, or robustness (cf. Sect. 2.2), to analyze the simulation results. In fact, the core positions, whose evolutions we have studied so far, possess, qua partial positions, a specific degree of justification which changes as the dialectal structure grows. In the following, we focus on the proponent core positions' robustness at an early stage of the debate, namely, at the density $D = 0.15$. And we distinguish proponent core positions (1) with a relatively high and (2) with a

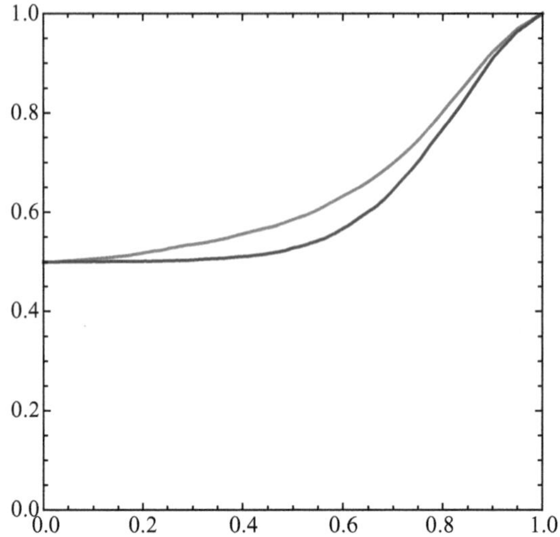

Fig. 8.2 Ensemble-wide normalized mean agreement of core proponent positions as a function of inferential density. Average normalized agreement among the proponents' cores is plotted for this chapter's ensemble with *lexicographic closest coherent* update (*bottom curve*) and for the ensemble, presented in Chap. 4, with simple *closest coherent* (*top curve*). In this second case, the proponents' core beliefs are, again, presumed to relate to the very same five sentences which make up the proponents' cores in the first ensemble (see Fig. 8.1)

Fig. 8.3 Ensemble-wide mean number of nonidentical core positions as a function of inferential density. The number of nonidentical cores is plotted for this chapter's ensemble with *lexicographic closest coherent* update (*a*) and for the ensemble, presented in Chap. 4, with simple *closest coherent* (*b*). Once more, the proponents' core beliefs in the second ensemble are presumed to relate to the very same five sentences which make up the proponents' cores in the first ensemble (see Fig. 8.1)

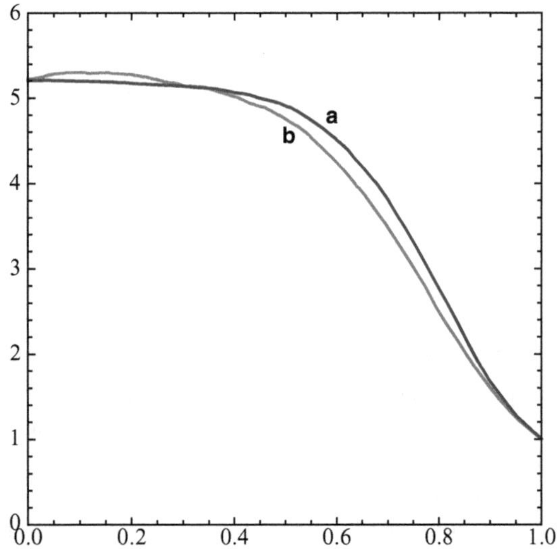

comparatively low robustness. High-robustness cores fall into the upper quartile of all cores in the ensemble; low-robustness cores belong, accordingly, to the lower quartile. Figure 8.4 displays how those proponent cores, with high respectively low robustness at an early stage, evolve in the subsequent debate as compared

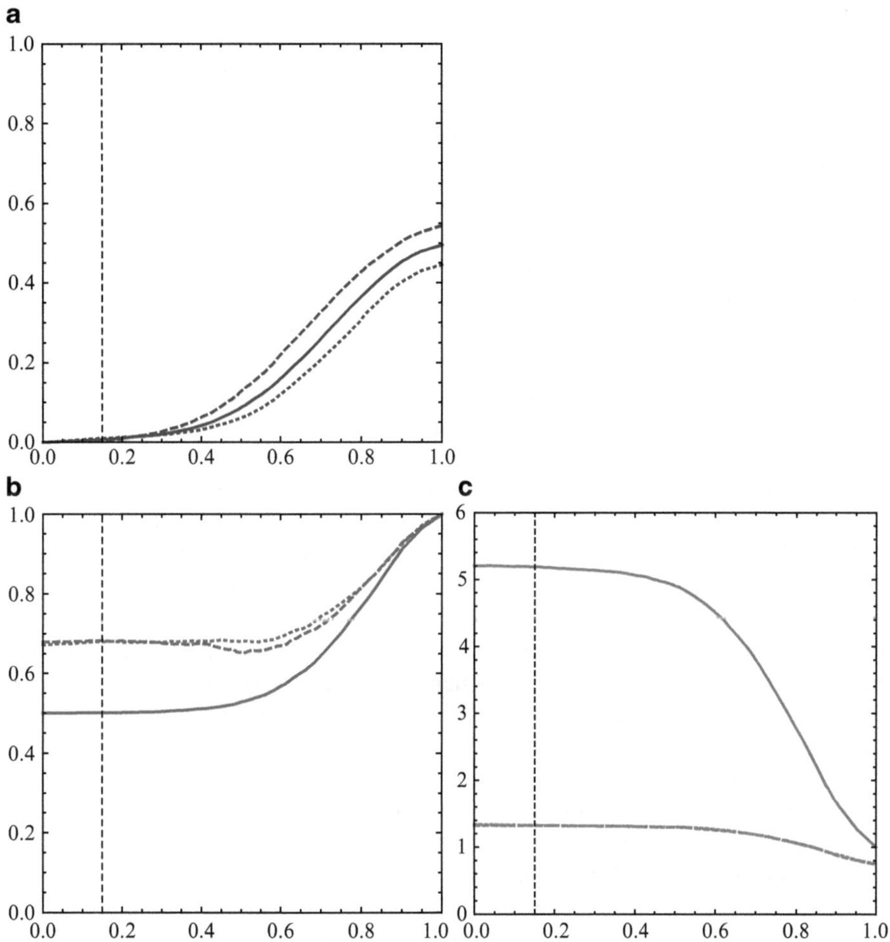

Fig. 8.4 Ensemble-wide mean normalized distance between actual and initial core proponent positions (**a**), ensemble-wide mean normalized agreement of core proponent positions (**b**), and ensemble-wide mean number of nonidentical core positions (**c**)—all plotted as functions of inferential density. The plots display ensemble-wide means as averaged over all proponents (*solid curves*), proponents with a very robust core position at $D = 0.15$ (*dotted curves*), and proponents who hold a core position with very low robustness at $D = 0.15$ (*dashed curves*). More specifically, a partial core position with high (low) robustness possesses a degree of justification which falls in the upper (lower) quartile of all robustness scores at the corresponding density in the ensemble

to the ensemble-wide mean. The core position dynamics of proponents which, at $D = 0.15$, maintain a core with high (low) robustness is pictured by dotted (dashed) curves.

As Fig. 8.4a shows, the degree of robustness at an early stage of the debate has a significant influence on how far and how rapidly a proponent's core recedes, subsequently, from its original position. Proponents with low core robustness withdraw more rapidly from their initial point of view than the average proponent does. Proponents with highly robust cores, however, give in at a lower pace. In the end, the final position attained by proponents with robust (unrobust) cores exhibits above-average (below-average) agreement with their initial position.

Figure 8.4b plots the mean agreement among proponents with high or, respectively, low robustness (at $D = 0.15$) and compares it to the mean agreement among all the proponents in a debate. In general, proponents with extreme robustness values exhibit fairly similar agreement evolutions. Surprisingly, initial agreement between proponents who occupy core positions with high or, respectively, low robustness equals roughly 67% and is much higher than the average initial agreement in the ensemble. As the density increases, agreement stays at this level and starts to rise only beyond a density of 0.6. In the density interval 0.4–0.8, agreement among proponents with low-robustness cores is marginally smaller than among those with highly robust cores.

The evolution of the number of nonidentical core positions is, eventually, shown in Fig. 8.4c. Exhibiting even greater similarity than in plot (b), the curves for high and low robustness have now become virtually indistinguishable. Yet, unlike in the case of mean agreement, it is not at all astonishing that the number of nonidentical core positions, maintained by proponents with high (or low) robustness at $D = 0.15$, lies way below the ensemble average. For this simply reflects the fact that, on average, only some of the proponents adopt relatively robust (unrobust) core positions. The number of nonidentical cores stays, on average, constant at almost 1.5 for the good part of a debate and starts its eventual decline at $D = 0.6$. As in the case of mean agreement, the dynamics of cores with extreme robustness thus mimics, at least roughly, the ensemble-wide mean evolution.

We've learned from Fig. 8.4a that robust core proponent positions tend to be more stable in the subsequent debate and evolve, on average, into a final position with a comparatively low distance to the corresponding initial position. In other words, the more robust a core position—even at a very early stage of the debate—the closer it already lies to the eventual consensus, or so it seems. Figure 8.5 attempts to spell out this result in some more detail. It affirms our previous finding: Core positions with a high robustness at an early stage of the debate tend to be slightly closer to the final position, closer to the consensus reached in the debate. More specifically, the positive relation between robustness and durability of a core position appears to be more pronounced in the left-hand plot, where the *reference position* is equated with the position reached when the number of nonidentical cores drops below 3 (as Fig. 8.3 informs us, this happens typically at $D \approx 0.8$), as opposed to the right-hand plot which considers, more accurately, the full consensus as reference position. Yet, in both plots proponent cores with very low robustness at $D = 0.15$ display, on average, a greater distance to the final position than cores with higher robustness.

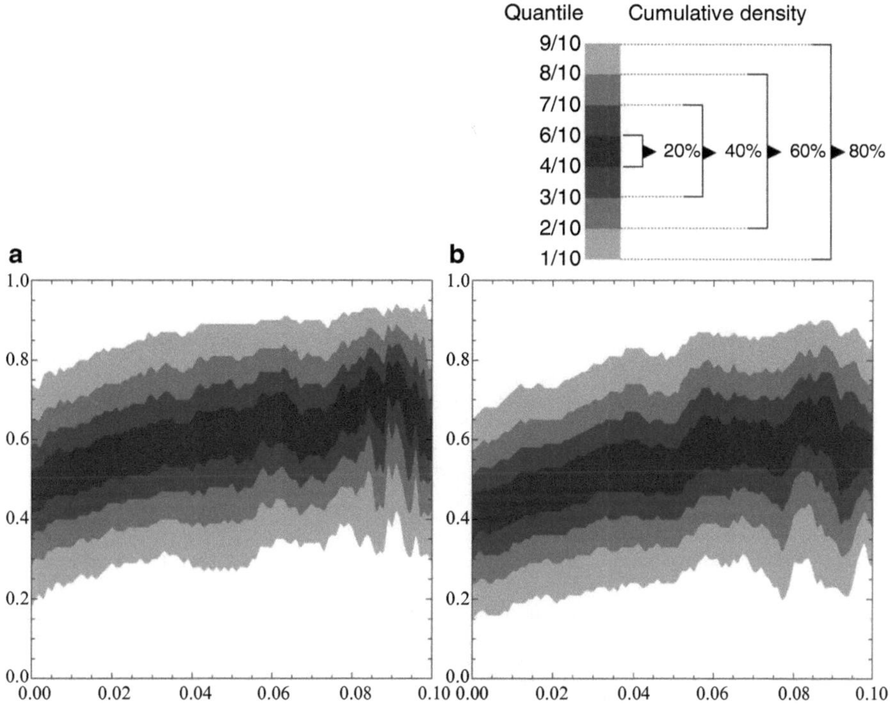

Fig. 8.5 Durability of core positions as a function of their robustness. The durability is measured by the mean normalized agreement of the proponent's core position at density $D = 0.15$ with a specific *reference position* in the corresponding debate. In panel (**a**), reference positions are presumed to consist in those positions the proponents hold once the number of nonidentical cores drops below 3. In panel (**b**), the full consensus position eventually attained represents the reference position. The shadings in the fan chart indicate the different quantiles as specified in the legend. The quantiles are calculated as follows: For each robustness value, a smooth probability density function (PDF) is fitted to the discrete relative frequencies of different durability values. This interpolated PDF is then used to derive the quantiles

8.3 Discussion

Two observations we have made in the previous section deserve a more detailed discussion:

1. The initial agreement among partial positions with extreme degrees of robustness (at $D = 0.15$) largely exceeds the ensemble's mean initial agreement (equal to 0.5).
2. The agreement between the core position a proponent holds at an early stage of the debate and the debate's final consensus positively depends on the core position's degree of justification.

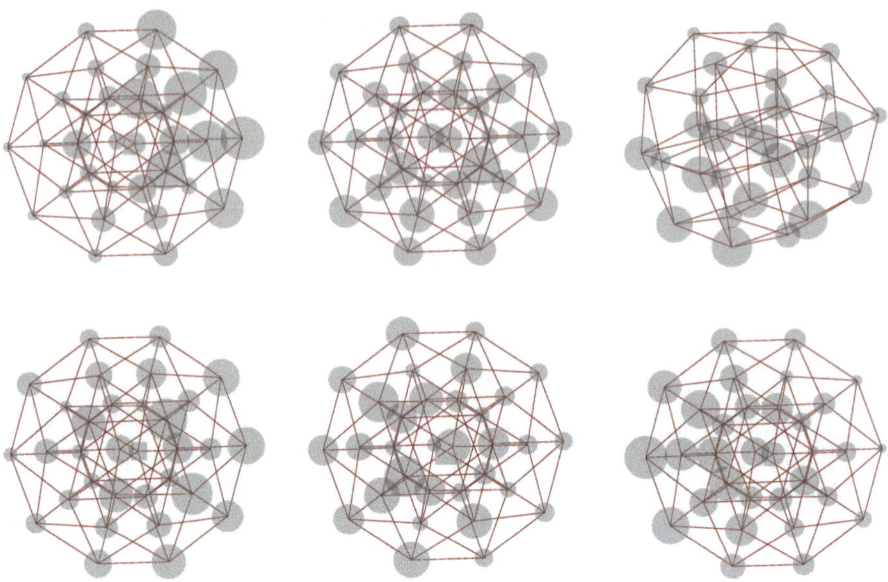

Fig. 8.6 Robustness of coherent core positions in six debates of this chapter's ensemble. The plots display the 5-dimensional section of the corresponding debates' SCP, at $D = 0.15$, taking into account sentences that belong to the proponents' cores only. The volume of the *circles* indicates the corresponding cores' robustness

The first observation suggests that closely related partial positions exhibit similar robustness. In other words, neighbors of robust positions tend to be robust, and neighbors of nonrobust positions tend to be nonrobust. This hypothesis would, in any case, explain why partitioning all proponent cores into robust and nonrobust ones yields two subsamples which exhibit significantly higher internal agreement than the entire set of partial proponent positions. Our hypothesis receives some initial support by Fig. 8.6. Its graphs depict the coherent core positions at a density of 0.15 for an illustrative sample of debates. As indicated by the circles' size, robust partial positions on the one side and nonrobust ones on the other side gather in different parts of the SCP. A water-tight validation, however, is provided by Fig. 8.7, which displays the correlation between a core's robustness and its neighbors' robustness. Obviously, the correlation is strong. Highly robust partial positions tend to be closely related to other robust positions. This fact, which neatly explains the high initial agreement between proponent cores with extreme degrees of justification, can itself be understood as a result of how ongoing argumentation shapes the space of coherent positions. Thus, as a first thing to note, the closer two partial positions, the more similar are their dialectic implications[1] regarding all the debate's sentences. But similarity of implications increases the probability that a

[1]That is their implications given the arguments uncovered and explicitly represented in the dialectical structure so far.

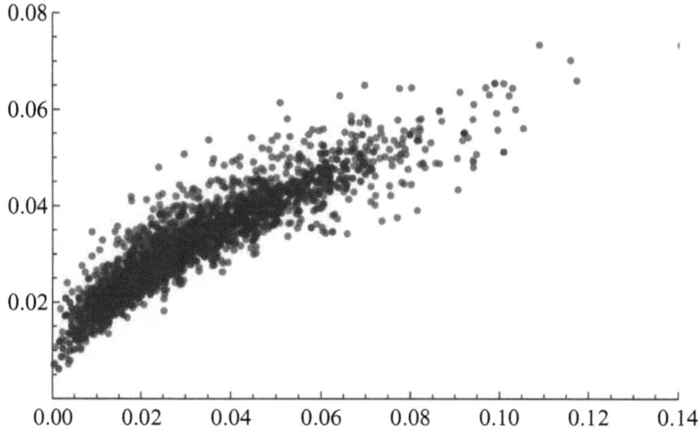

Fig. 8.7 Correlation between a core position's robustness and its neighbors' robustness. For each coherent core position $\mathscr{P}$ at $D = 0.15$, the core's robustness (x-axis) is plotted against the average robustness of its adjacent core positions (y-axis)—whereas an adjacent core position $\mathscr{P}'$ disagrees with $\mathscr{P}$ in regard to exactly one core sentence ($\mathrm{HD}(\mathscr{P}, \mathscr{P}') = 1$). The plot considers all coherent core positions in a random sample of 100 debates, drawn from this chapter's ensemble

given complete position represents either a coherent extension of both or of none of the two different partial positions. Or, to put it differently, if two positions possess rather disparate implications, a coherent extension of one of them is relatively unlikely to be a coherent extension of the other one as well. So, it becomes comprehensible that very similar partial positions possess, to a large degree, the same coherent extensions—and thence roughly the same degree of robustness.

Let us turn to the second observation. How come a core's robustness at an early stage of a debate determines the agreement of this very partial position with the debate's final position? Why do robust positions tend to be closer to the consensus proponents eventually settle on? In order to answer these questions, we have to step back for a moment and reconsider what a partial position's high degree of justification signifies—beyond the formal definition given in Sect. 2.2—in the first place. In the book *Theorie dialektischer Strukturen*, I suggest the following interpretation of degrees of justification (cf. Sect. 74). Degrees of justification indicate a partial position's prima facie ability to resist future falsifications. This is because a high degree of justification implies, by definition, that the partial position possesses many coherent extensions. If one of these complete positions, which extend the partial position, is rendered incoherent—by a newly introduced argument or the expansion of the background knowledge— the partial position itself can typically still be coherently adopted, for it possesses further coherent extensions. A core position with low robustness, in contrast, is very susceptible to falsification. Assume that it is extended by merely one complete and coherent position. If this complete position is rendered incoherent in the course of the debate, there is no way of holding on to the core position. Opinions outside the debate's core cannot be adjusted so as to retain the core beliefs. The partial position itself has to be given up.

I presume that it is this feature of the concept of degree of justification, which I conjectured somewhat vaguely in *Theorie dialektischer Strukturen*, that explains the observed positive correlation between a partial position's robustness at an early stage of the debate and its agreement with the final consensus. Robust positions are less susceptible to future falsifications, have to be modified less frequently during the subsequently unfolding debate, and therefore agree to a larger extent with the final consensus. The findings of the simulations presented in this chapter thus corroborate, in general, the analysis of degrees of justification in *Theorie dialektischer Strukturen* and demonstrate, in particular, why striving to maximize the robustness of one's core position represents a rational discursive aim.

Chapter 9
The Consensual Dynamics of Debates with Core Argumentation

In the previous chapter, we distinguished between core beliefs, which proponents give up only reluctantly, and auxiliary beliefs, beyond the debate's core, which proponents are much more willing to alter. This distinction translated into a modified update mechanism, namely, the *lexicographic closest coherent* updating. We have studied the effect of this new updating procedure while retaining the simple *random argumentation* mechanism. Clearly, core beliefs can also be taken into account when putting forward new arguments. Thus, we may devise various argumentation mechanisms which are sensitive to the distinction between core and auxiliary beliefs. In this chapter, we examine two such mechanisms. The first one is derived from the most effective argumentation strategy studied so far. The *multiple core convert* strategy tries to convert as many opponents as possible while explicitly targeting their core convictions. The design of the second argumentation mechanism we consider in this chapter is motivated by our previous observation that a core's robustness exerts a significant influence upon the future evolution of the proponent's position. This suggests to maximize, as an argumentation rule, the robustness of one's core position.

9.1 Setup

For each of the two argumentation mechanisms, we set up an ensemble of 1,000 debates with 6 proponents each. The simulations terminate as soon as the proponents agree on a core position. Let $\mathscr{P}_t^i$ denote the core position proponent i holds at step t, and let $\mathscr{Q}_t^i$ refer to her corresponding complete position, which extends $\mathscr{P}_t^i$. In the first ensemble, proponents apply the *multiple core convert* strategy, that is:

Argumentation Mechanism: The proponents introduce, in successive order, new arguments in line with the following rule. The proponent i chooses, randomly, one core sentence she considers true—formally, $c \in \{s \in S | \mathscr{P}_t^i(s) = \text{true}\}$. This sentence c makes up the conclusion of the new argument. She identifies,

G. Betz, *Debate Dynamics: How Controversy Improves Our Beliefs*,
Synthese Library 357, DOI 10.1007/978-94-007-4599-5_9,

subsequently, two further different sentences, $p_1, p_2 \in S$, from the entire sentence pool so as to maximize the number of opponents who hold both sentences. The choice of p_1, p_2 hence maximizes $|\{j|\mathscr{P}_t^j(p_1) = \mathscr{P}_t^j(p_2) = \text{true}\}|$. These two sentences constitute the argument's premisses—provided that introducing the new argument leaves at least one position coherent (otherwise, the proponent chooses, under the same maximality condition, a different pair of premisses). Adding $(p_1,p_2;c)$ to τ_t yields τ_{t+1}. We call this argumentation strategy *multiple core convert*.

Discovery Mechanism: There is no background knowledge.

Update Mechanism: *Lexicographic closest coherent* (cf. Sect. 8.1).

The debates in the second ensemble are set up as follows:

Argumentation Mechanism: The proponents put forward, in alternating sequence, one argument each. The new argument is drawn from the set of all potential arguments, that is, noncircular arguments which leave at least one position coherent, so as to maximize the robustness of the corresponding proponent's core position.[1] This argumentation mechanism will be referred to as *robust argumentation*.

Discovery Mechanism: There is no background knowledge.

Update Mechanism: *Lexicographic closest coherent* (cf. Sect. 8.1).

9.2 Results

In random debates with lexicographic update mechanism, studied in the previous chapter, proponents modify their core beliefs only very reluctantly (cf. Fig. 8.1). Such doxastic inertia, however, is overcome by employing *multiple core convert* and *robust argumentation*. As Fig. 9.1 shows, these purposeful and core-aware argumentation strategies compel the debates' proponents to retreat from their initial positions more rapidly as compared to debates with *random argumentation*. Moreover, *multiple core convert* (a) is substantially more effective in forcing proponents to modify their core beliefs than *robust argumentation* (b). Accordingly, proponents disagree, at the density $D = 0.5$, with 55% of their initial core convictions when employing *multiple core convert*, as compared to less than 30% when resorting to *robust argumentation*. The "conservative" update mechanism *lexicographic closest coherent* doesn't prevent the drastic modification of core positions. Surprisingly, proponents eventually deviate from their initial core position by more than 50% with *multiple core convert*, that is, the final consensus in debates with *multiple core convert* is, on average, more distant from the initial proponent positions than a randomly selected position. Both with *random* and with *robust argumentation*, in contrast, initial proponent positions agree with half the final consensus' statements.

[1] Due to limited computational resources, proponents actually consider at maximum 100 different potential arguments.

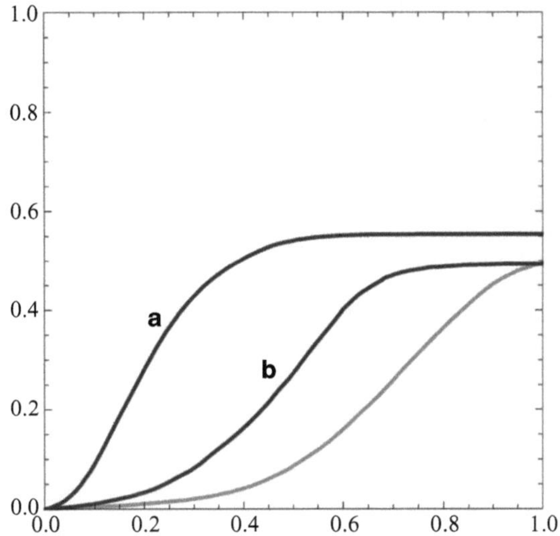

Fig. 9.1 Ensemble-wide mean normalized distance between current and initial core proponent position as a function of inferential density. Mean distance to initial position is plotted for this chapter's ensembles with *multiple core convert* (**a**) and *robust argumentation* (**b**) as well as for the ensemble, presented in Chap. 8, with *random argumentation* (*right-hand curve, light gray*)

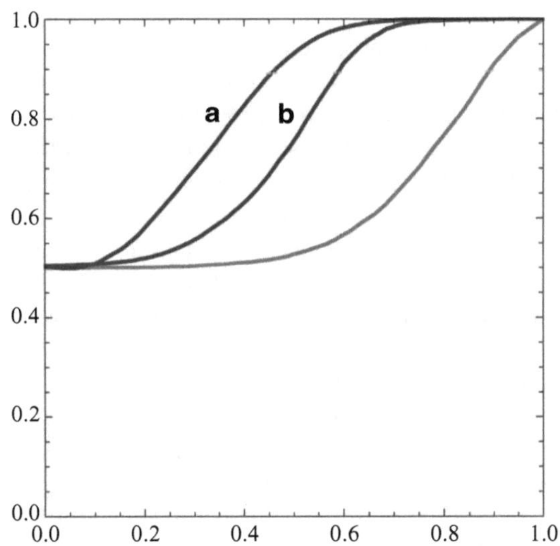

Fig. 9.2 Ensemble-wide mean normalized agreement of core proponent positions as a function of inferential density. Average normalized agreement among the proponents' cores is plotted for this chapter's ensembles with *multiple core convert* (**a**) and *robust argumentation* (**b**) as well as for the ensemble, presented in Chap. 8, with *random argumentation* (*right-hand curve, light gray*)

Figure 9.2a, plotting mean agreement evolutions, suggests that the substantial modifications of core beliefs in the *multiple core convert* ensemble are the price proponents pay in order to achieve rapid rapprochement in spite of the conservative update mechanism. With *multiple core convert*, mean agreement increases not only much more quickly than with *random argumentation* but also more rapidly than with *robust argumentation*. At $D = 0.5$, when core positions have hardly approached each other in the *random argumentation* ensemble, mean agreement has increased from 0.5 to 0.75 (*robust argumentation*) and, respectively, to almost 0.95 (*multiple core*

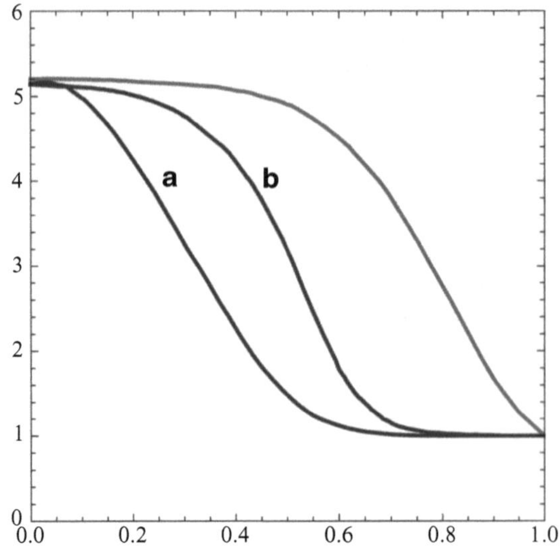

Fig. 9.3 Ensemble-wide mean number of nonidentical core positions as a function of inferential density. The evolution of nonidentical cores is plotted for this chapter's ensembles with *multiple core convert* (**a**) and *robust argumentation* (**b**) as well as for the ensemble, presented in Chap. 8, with *random argumentation* (*right-hand curve, light gray*)

convert). Thus, argumentation strategies exert a considerable influence not only on how fast the proponents' complete positions (cf. Chaps. 6 and 7) but also on how fast their core positions approach each other.

The superior consensus-conduciveness of this chapter's argumentation strategies is also reflected in the evolution of the number of nonidentical core positions (Fig. 9.3). As a consequence of the random sampling, there are initially roughly five nonidentical proponent core positions per debate. With *random argumentation*, this hardly changes up to a density of $D = 0.5$. During this phase of a debate, however, the number of nonidentical core positions drops to roughly 3 with *robust argumentation* (b) and to 1.5 with *multiple core convert* (a). Hence, *multiple core convert* engineers effectively full core consensus.

We shall now consider the impact of a core's robustness, that is, its degree of justification, on the position dynamics. As in the previous chapter, we take $D = 0.15$ as the reference density and calculate the degrees of justification of the proponent core positions at this specific point. Henceforth, we distinguish proponent positions which fall, in regard of these degrees of justification, into the upper quartile (high robustness) and into the lower quartile (low robustness). Figure 9.4 gauges the pace at which highly robust (dotted curves) and unrobust (dashed curves) proponent core positions retreat from the corresponding initial positions, as compared to the ensemble-wide mean. In the ensemble with *multiple core convert* (left-hand panel), proponent cores with high robustness have, at $D = 0.15$, modified substantially more individual beliefs (roughly 30%) than cores with low robustness (ca. 10%). In the long run, however, this difference vanishes: Proponents of both types will eventually disagree with roughly 55% of their initial core positions.[2] With *robust*

[2]This result is robust against decreasing the reference density, for example, to 0.05.

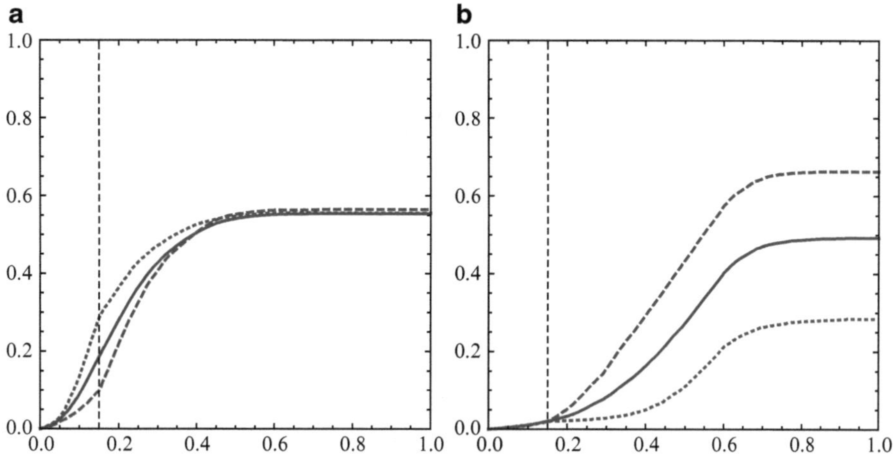

Fig. 9.4 Ensemble-wide mean normalized distance between current and initial core proponent positions as a function of inferential density, plotted for this chapter's ensembles with *multiple core convert* (**a**) and *robust argumentation* (**b**). The plots display ensemble-wide means as averaged over all proponents (*solid curves*), over proponents with a very robust core position at $D = 0.15$ (*dotted curves*), and over proponents who hold a core position with very low robustness at $D = 0.15$ (*dashed curves*). More specifically, a partial core position with high (low) robustness possesses a degree of justification which falls in the upper (lower) quartile of all robustness scores at the corresponding density in the ensemble

argumentation (right-hand panel), however, things look different. At $D = 0.15$, the proponent core positions, no matter whether robust or not, have barely been modified at all. But at higher densities, proponents who hold a very robust core position on the one side and proponents whose core position displays a low degree of justification (at $D = 0.15$) on the other side start to diverge significantly. Proponents with robust core positions hardly change their beliefs until a density of 0.4 is reached. And even in the long run, they agree with their corresponding initial position to a much larger extent than the ensemble-wide mean. Proponents whose core position possesses a comparatively low degree of justification (at $D = 0.15$), however, begin to withdraw from their starting point once the density of 0.15 has been passed. Eventually, these proponents modify more than 65% of their initial beliefs.

Figure 9.5 plots the mean agreement between proponents with robust and, respectively, unrobust core positions. Initially, as well as at $D = 0.15$, proponent positions with extreme robustness values exhibit above-average mutual agreement. Yet—and this marks a difference to *random argumentation* (cf. Fig. 8.4)—proponents with highly robust cores agree, initially as well as throughout the debate, to a significantly greater extent than proponents whose cores exhibit low robustness (at $D = 0.15$). This applies to both ensembles. Moreover, in the *multiple core convert* ensemble, mean agreement among proponents with robust cores soars at very low densities, increasing from 65% at $D = 0$ to almost 80% at $D = 0.15$.

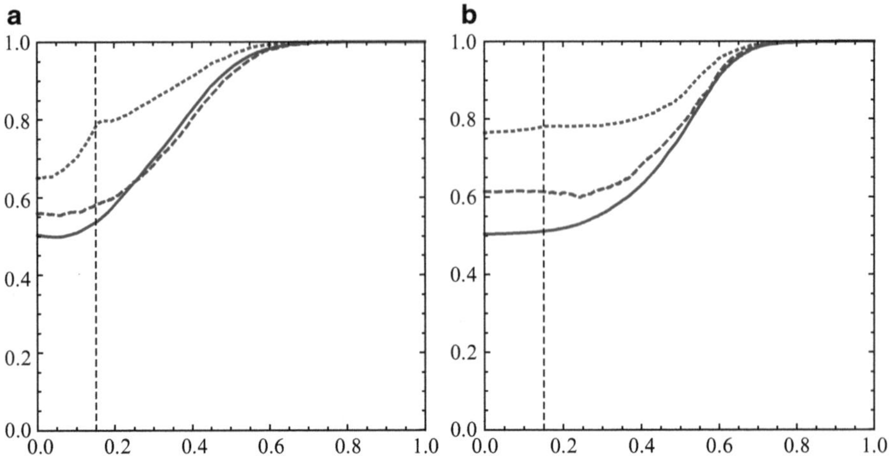

Fig. 9.5 Ensemble-wide normalized mean agreement of core proponent positions as a function of inferential density, plotted for this chapter's ensembles with *multiple core convert* (**a**) and *robust argumentation* (**b**). See Fig. 9.4 for further information

This augmentation concurs with the fact that the corresponding proponents' core positions undergo substantial, early modifications, as we have found above (see Fig. 9.4a). Except for this peculiarity, mean agreement among proponents with very robust and, respectively, unrobust partial positions seems to increase approximately in tune with overall agreement.

The evolution of the number of nonidentical core positions, averaged over proponents whose cores display extreme degrees of robustness at $D = 0.15$, is shown in Fig. 9.6. Unsurprisingly, these curves lie well below the ensemble-wide mean, since, typically, only a few of the debate's proponents hold partial positions with very high or very low degree of justification. In the ensemble with *multiple core convert* argumentation (left-hand panel), a higher proportion of proponents which maintain a robust (as compared to a nonrobust) position at $D = 0.15$ reach full consensus at densities between 0.1 and 0.5. In the ensemble with *robust argumentation* (right-hand panel), however, the numbers of nonidentical core positions held by the two types of proponents evolve in virtually identical ways.

Reconsider Fig. 9.4, which depicts how far proponents retreat from their doxastic starting point. Now, the less a proponent has to withdraw from her initial position in the course of a debate, the more her initial position agrees with the final consensus position. Moreover, core positions with high and low degrees of justification at $D = 0.15$ are, in the light of the subsequently introduced arguments, subject to modifications of varying extent. This suggests, like in the previous chapter, that robustness might indicate proximity to a debate's eventual consensus. Particularly in the ensemble with *robust argumentation*, where the modifications differ substantially (Fig. 9.4b), degrees of justification presumably constitute a reliable indicator

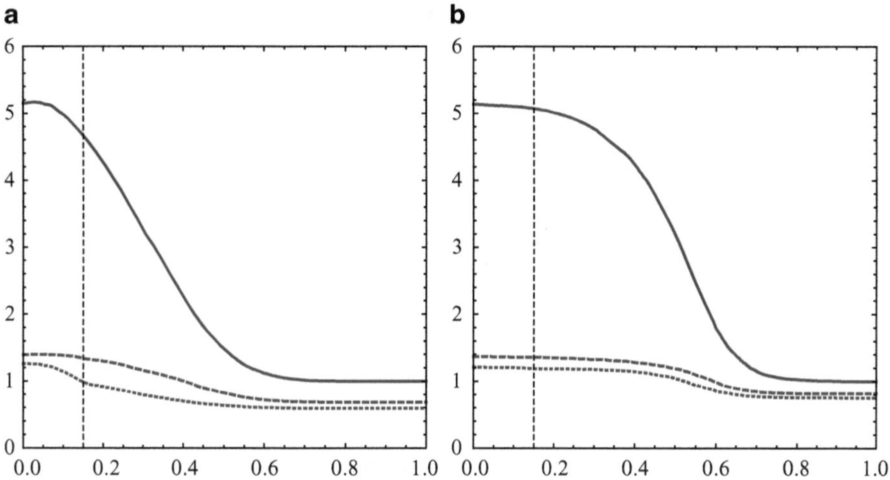

Fig. 9.6 Ensemble-wide mean number of nonidentical core positions as a function of inferential density, plotted for this chapter's ensembles with *multiple core convert* (**a**) and *robust argumentation* (**b**). See Fig. 9.4 for further information

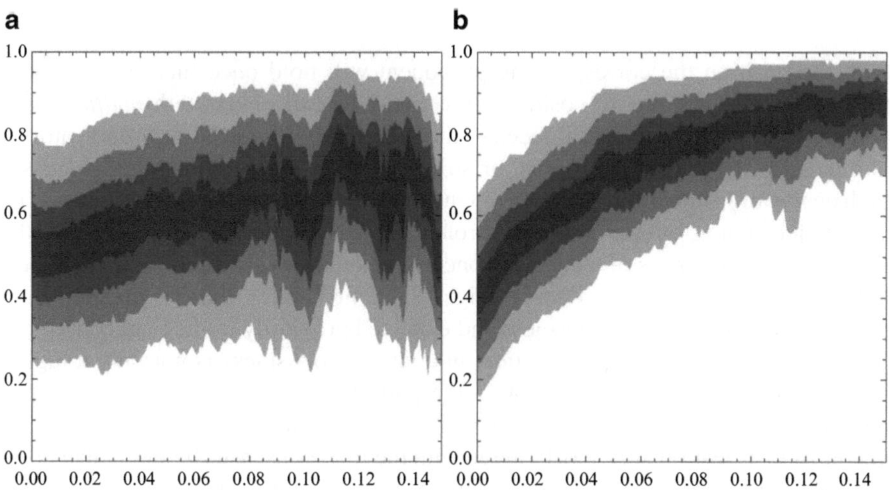

Fig. 9.7 Durability of core positions as a function of their robustness. The durability is measured by the mean normalized agreement of the proponent's core position at density $D = 0.15$ with the position she holds once the number of nonidentical proponent cores drops below 3. Panel (**a**): Ensemble with *multiple core convert*. Panel (**b**): Ensemble with *random argumentation*. For further details, compare Fig. 8.5

of consensus proximity. As Figs. 9.7 and 9.8 demonstrate, this is the case, indeed. In the ensemble with *robust argumentation*, the robustness of a partial position at $D = 0.15$ is a highly accurate indicator of that position's agreement with the debate's future partial consensus—more precisely: an indicator of the agreement

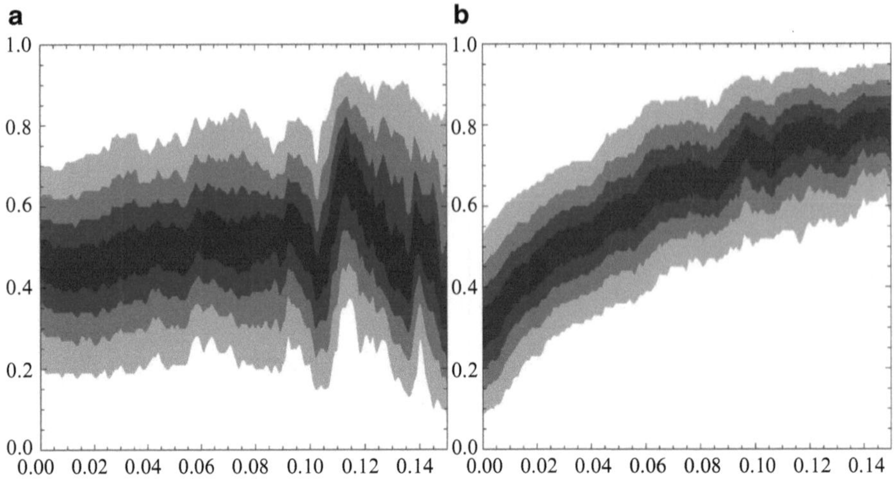

Fig. 9.8 Durability of core positions as a function of their robustness. The durability is measured by the mean normalized agreement of the proponent's core position at the density $D = 0.15$ with the debate's final consensus position. Panel (**a**): Ensemble with *multiple core convert*. Panel (**b**): Ensemble with *robust argumentation*. See also Fig. 8.5

with the position the corresponding proponent will hold once there are no more than two nonidentical core positions (see Fig. 9.7). Still, even with *multiple core convert*, robustness remains a decent—though clearly less accurate—indicator of closeness to the partial consensus, or so it seems[3]: Core positions that possess a high degree of justification at $D = 0.15$ tend to be closer to the partial consensus. The crucial question is, of course, whether robustness also indicates agreement with the final and full consensus of a debate. Concerning debates with *multiple core convert* argumentation, the answer is no. As Fig. 9.8a demonstrates, there is no significant positive relation between robustness and expected proximity to the final consensus. In debates with *robust argumentation*, in contrast, robustness is not only a highly revealing indicator of agreement with the partial but even with the eventual full consensus (Fig. 9.8b). Accordingly, 3/5 of the proponent core positions with a high degree of justification at $D = 0.15$ agree with the final consensus by more than 60%, and half of them even accord with at least 80% of the final consensus' statements. In contrast, half of the partial positions with very low robustness agree with the final consensus, on average, by less than 40%.

[3]The fluctuations at high degrees of justification indicate that the sample is not sufficiently large: There are only relatively few very robust core positions. They dominate the overall picture (at high degrees of justification) and may introduce a bias in case they exhibit, contingently, extreme durability values.

9.3 Discussion

We will submit the following facts, reported above, to a more detailed discussion in this section:

1. The purposive argumentation mechanisms studied in this chapter display a substantially higher consensus-conduciveness—even with regard to (reluctantly revised) core beliefs—than the simple *random argumentation* which we have investigated in the previous chapter.
2. In debates with *multiple core convert*, proponent core positions approach each other at a higher pace than in *robust argumentation* debates.
3. Proponents with highly robust positions exhibit, initially, greater mutual agreement than proponents whose core positions possess an extremely low degree of justification (cf. Figs. 9.5 and 9.6).
4. In the ensemble with *multiple core convert*, proponents with highly robust positions have, at $D = 0.15$, withdrawn substantially from their corresponding starting points.
5. With *robust argumentation*, there is a strong correlation between a core position's robustness at a very early stage of the debate and its expected proximity to the debate's final consensus. With *multiple core convert*, however, robustness of a proponent's core merely seems to indicate the distance to a future partial consensus, more specifically the agreement with the core position the proponent holds once there are no more than two nonidentical proponent core positions. But we find, regarding *multiple core convert*, virtually no positive relationship between robustness and expected agreement with a debate's final and full consensus.

Consider the first fact in our list. It doesn't come as a surprise, given the results of previous chapters, that the modified *multiple convert* argumentation rule is more consensus-conducive—even with respect to core sentences—than the simple *random argumentation*. In order to see why *robust argumentation*, too, leads to faster rapprochement, we consider its logic on a simple and illustrative space of coherent positions which is defined on a pool of three sentences (p_1, p_2, p_3). Let's assume that every combinatorially possible truth-value assignment represents a coherent position and that a given proponent considers the three sentences true, with $[p_1]$ being her core position. Figure 9.9a depicts all complete coherent positions which extend the proponent's core. The number of positions that extend the core divided by the number of all coherent positions gives the degree of justification of the partial position $[p_1]$. It thus equals 1/2. Panel (b) depicts the situation where both p_1 and p_2 represent core sentences. There are, accordingly, only two complete and coherent positions which extend the proponent's core. Its degree of justification equals, consequently, 1/4.

An argumentation strategy that maximizes the robustness of a proponent core position attempts to render complete positions which don't extend the corresponding partial position incoherent while leaving complete positions which do extend it

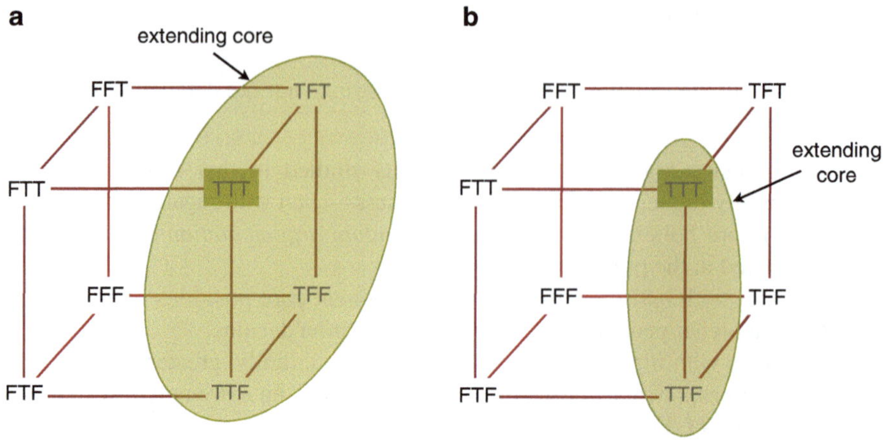

Fig. 9.9 Complete coherent positions which extend a proponent's core position in an illustrative space of coherent positions (see also Fig. 6.5). Panel (**a**): p_1 is the only core sentence; accordingly, the proponent core position consists in assigning p_1 the value *true*. This partial position is extended by all coherent positions which claim that p_1 is true (*khaki circle*), including the complete proponent position itself (*khaki rectangle*). Panel (**b**): In contrast to panel (**a**), both p_1 and p_2 are core sentences. The proponent core hence consists in $[p_1, p_2]$ and is extended by two complete coherent positions

intact. In terms of Fig. 9.9a, arguments that maximize robustness target, primarily, the coherent positions outside the khaki circle. Each of these four positions is rendered incoherent by one of the following arguments: $(p_2,p_3;p_1)$, $(\neg p_2,p_3;p_1)$, $(p_2,\neg p_3;p_1)$, and $(\neg p_2,\neg p_3;p_1)$. Introducing exactly one of these arguments increases the robustness from 1/2 to 4/7; introducing all arguments increases it to 1. Figure 9.10a highlights the complete positions these four arguments render incoherent (gray); the positions which extend the proponent's core position (on the right-hand side of the cube) remain coherent. Still, we should note that an argument may increase the robustness of a core position in spite of eliminating coherent positions which extend the core, as the following example demonstrates. In Fig. 9.10b, an argument which derives p_2 from the background knowledge is introduced into a debate. As some complete positions had been rendered incoherent before, this argument renders three previously coherent positions incoherent, only two of which don't extend the proponent's core. As a consequence, the new argument increases the core's degree of justification from 1/2 to 2/3.

In sum, by following the *robust argumentation* rule, a proponent tries to eliminate as many coherent positions which disagree with her core beliefs as possible. Accordingly, she targets positions which are, actually or potentially, held by opponents who disagree with her core convictions. Opponents that share the proponent's core beliefs, however, are addressed and forced to alter their positions to a much lesser extent. This explains why *robust argumentation* generates core consensus more effectively than the pure *random argumentation*.

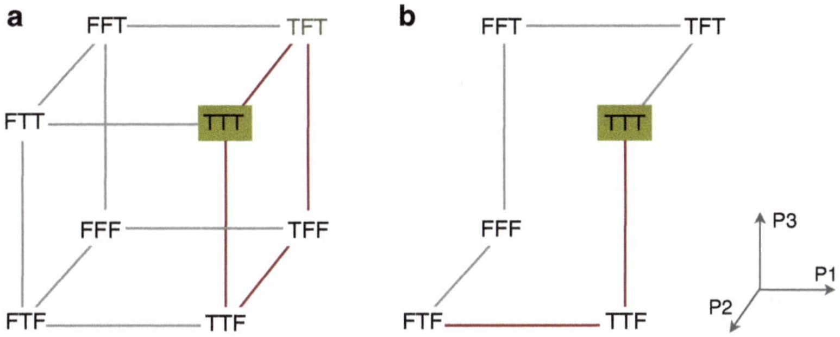

Fig. 9.10 Positions rendered incoherent by introducing arguments so as to increase the robustness of the proponent's core. In both illustrations, p_1 represents the only core sentence (see also Figs. 6.5 and 9.9a). Panel (**a**): Complete positions which are rendered incoherent by one of the arguments $(p_2,p_3;p_1)$, $(\neg p_2,p_3;p_1)$, $(p_2,\neg p_3;p_1)$, and $(\neg p_2,\neg p_3;p_1)$ are colored *gray*. They don't extend the proponent's core and therefore increase its degree of justification. Panel (**b**): Positions which are rendered incoherent by introducing an argument that derives p_2 from the (implicit) background knowledge. Because some complete positions had been rendered incoherent before (gaps), the new argument increases the core's robustness although it falsifies positions which extend the core

But why is it that *multiple core convert* is significantly more consensus-conducive than *robust argumentation*, although the latter directly targets core dissent? This brings us to the second observation to be explained. I suggest that, in this regard, the crucial feature of *multiple core convert* consists in its attacking or supporting but core theses. The conclusion of an argument which is introduced in line with *multiple core convert* is always a core sentence. In addition, such an argument may recruit its premisses from the core sentences as well. As a consequence, *multiple core convert* spawns particularly dense inferential relations between the core sentences, leading to an above-average contraction of the corresponding section of the SCP. We shall introduce the concept of core inferential density so as to quantify this effect. Core inferential density is defined in close analogy to a debate's inferential density; yet, instead of *all* sentences and *all* coherent positions, it merely considers core sentences and the number of coherent partial positions defined on this core.[4] Figure 9.11 plots the evolution of core inferential density as a function of inferential density for the ensembles studied in this chapter. As this figure demonstrates, core density increases significantly faster with *multiple core convert* than with *robust argumentation*. *Multiple core convert* renders substantially more partial positions, defined on the core sentences, incoherent than *robust argumentation*, in particular at densities lower than 0.6. More specifically, the evolutions of core densities in the two

[4]Formally, the core density of a dialectical structure τ with a sentence pool of size $2n$ and $2n'$ core sentences ($n' < n$) equals

$$D_{\text{core}}(\tau) = \frac{n' - \lg(\sigma'_\tau)}{n'},$$

where σ'_τ denotes the number of coherent partial positions defined on the $2n'$ core sentences.

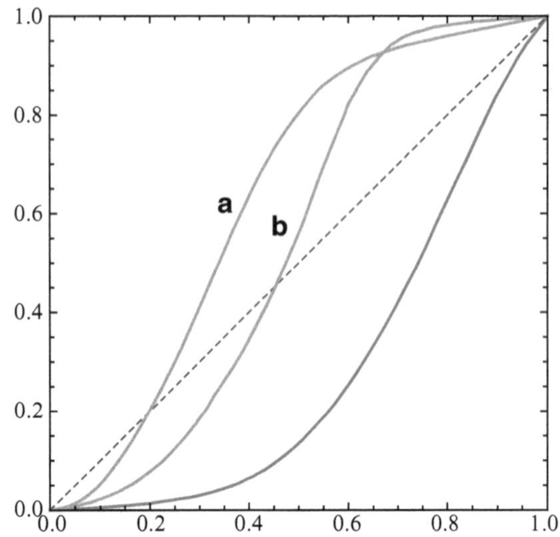

Fig. 9.11 Ensemble-wide mean core density as a function of inferential density. Average core density (y-axis) is plotted for this chapter's ensembles with *multiple core convert* (**a**) and *robust argumentation* (**b**) as well as for the ensemble, presented in Chap. 8, with *random argumentation* (*right-hand curve*)

ensembles dovetail nicely with the evolutions of mean core agreement (Fig. 9.2). The more rapid increase of core density therefore provides a good explanation for the superior core consensus-conduciveness of *multiple core convert.*

The third observation enumerated above suggests that, both in the ensemble with *multiple core convert* and in the ensemble with *robust argumentation*, the spatial correlation (in the SCP) between highly robust core positions is significantly greater than the spatial correlation between core positions with low degree of justification. This can be understood by a spillover effect: The two argumentation strategies tend to increase the robustness of core positions in the vicinity of the proponent who sets forth an argument. As a consequence, if two proponents initially agree, by chance, to a high degree, they are much more likely to attain a very robust core position (at $D = 0.15$) because they mutually "benefit," in terms of increasing robustness, from each other's argumentation. And if three or four proponents agree, that's even more beneficial. In contrast, the core position of an isolated proponent is typically only rendered more robust by that proponent's own and not by her opponents' argumentation (only every sixth and not every third or every second argument fosters her core position's robustness). Moreover, this spillover effect, I posit, allows proponents to attain substantially higher degrees of justification in this chapter's ensembles as compared to the ensemble with *random argumentation* (cf. Fig. 9.12). Let us explain in some more detail why precisely a proponent who introduces an argument in line with (1) the *multiple core convert* strategy or (2) the *robust argumentation* strategy tends to increase the robustness not only of her but also of further core positions nearby. Ad (1), we may note that a proponent, by following the *multiple core convert* strategy, supports one of her core beliefs with every new argument she introduces. In the same time, she supports, obviously, a core belief of every opponent who shares that very single conviction. And clearly, opponents who agree to a large extent with the proponent's overall core position

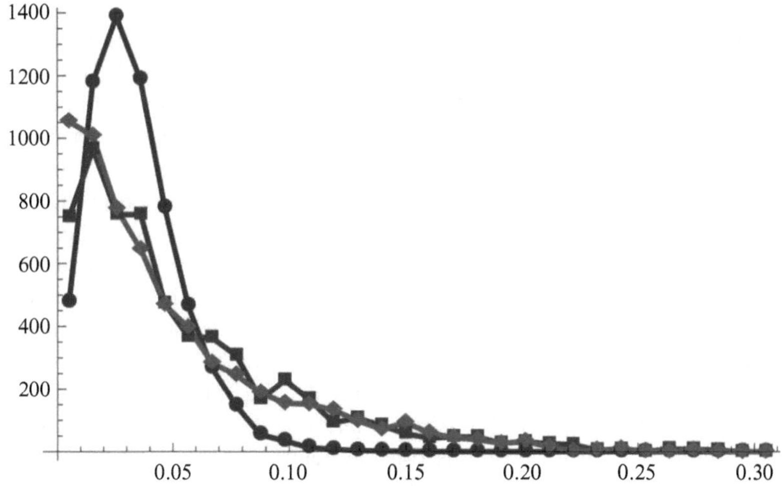

Fig. 9.12 Absolute frequencies of proponent core positions with corresponding robustness values (x-axis) at $D = 0.15$. Frequency distributions are plotted for this chapter's ensembles with *multiple core convert* (*squares*) and *robust argumentation* (*diamonds*) as well as for the ensemble, presented in Chap. 8, with *random argumentation* (*circles*)

are more likely to endorse that one particular core statement, too. Now, as has been argued in Betz (2011b), supporting a statement by a new argument tends to increase that very statement's degree of justification and hence, we may presume, fosters the robustness of the overall core position as well. Thus, by supporting their own as well as closely related core positions, proponents with high core agreement, who follow the *multiple core convert* strategy, mutually amplify their cores' robustness. Ad (2), it is helpful to reconsider the effects of *robust argumentation* on the SCP and their illustration in Figs. 9.9 and 9.10. Arguing so as to maximize a core position's robustness means, we have seen above, to eliminate complete positions from the SCP which don't extend the core while leaving positions which do intact. As already noted in Sect. 8.3, closely related core positions tend to have more complete extensions in common than partial positions which hardly agree. Therefore, if an argument leaves most complete positions that extend a given core position, $\mathcal{P}$, coherent, it is likely to leave many extensions of closely related core positions intact, too. Vice versa, in case the argument renders a complete position which does not extend $\mathcal{P}$ incoherent, the eliminated complete position is unlikely to be an extension of a core position in $\mathcal{P}$'s vicinity either. Consequently, maximizing the robustness of some core position through argumentation tends to increase the robustness of closely related core positions. Robustness growth spills over to nearby core positions, which explains why highly robust proponent core positions display, on average, substantial agreement.

Let us turn to the fourth observation. Why do proponents with highly robust cores retreat substantially from their initial positions in the *multiple core convert* ensemble? To see this, consider, first, the ensemble's debates which host the comparatively most robust core positions. A numerical analysis of the ensemble reveals that, in these debates, *all* proponent core positions—and not merely the highly robust ones—have retreated, at $D = 0.15$, on average by 0.28 from their initial position. So what is in need of explanation is not simply the fact that highly robust positions withdraw substantially from their starting point but that there exist some debates in which all proponents are compelled to modify a significant proportion of their initial beliefs, even at an early stage of the debate ($D = 0.15$). A further statistical analysis of the ensemble uncovers that highly robust core positions typically occur in debates which display a relatively high core density. This is quite easy to understand: Recall that, at a fixed density, for example, $D = 0.15$, debates accommodate the same number of complete and coherent positions. Under the assumption that all coherent core positions possess, at least roughly, the same number of complete and coherent extensions, it follows immediately that debates with relatively few remaining coherent partial positions (on the central sentences) contain the most robust proponent core positions. So high-robustness scores stem from high core density. In debates with high core density, however, a comparatively large proportion of proponent core positions, which were, initially, randomly distributed, have already been rendered incoherent and had to be modified, possibly several times. Thus, the higher the core density, the farther the proponents' cores are typically apart from their initial positions. That is the reason why, in the *multiple core convert* ensemble, very robust positions are, at $D = 0.15$, relatively distant from their starting point. In the ensemble with *robust argumentation*, however, this is apparently not the case. This difference can only spring from the fact that *robust argumentation* succeeds in generating core proponent positions with extreme robustness scores (cf. Fig. 9.12) while maintaining a relatively low core density at $D = 0.15$ (see Fig. 9.11).

The most relevant findings of this chapter pertain to the correlation between a core position's robustness at an early stage of the debate and its agreement with the debate's eventual final consensus. We've found evidence in favor of such a positive relationship in the previous chapter and succeeded in explaining it in terms of a core's resistance and immunity against future falsification. The picture has, however, become more complicated as we consider different argumentation rules in this chapter. The fifth and final observation, which summarizes our results, will be discussed in the following paragraphs.

As a first and at first glance disappointing result, we find that, in the *multiple core convert* ensemble, there is virtually no correlation between the robustness of a core position (at $D = 0.15$) and its expected agreement with the debate's final and full consensus. To understand this, we picture an extreme case: Consider an argumentation in the course of which every proponent core position is rendered incoherent with each new argument. The proponents are hence compelled to modify their core positions at every step. In such a case, the robustness of a proponent's core does obviously not tell anything about her future doxastic state. For no matter how robust the core position, the next argument will render it incoherent anyway, and the proponent will move to a (possibly radical) different new core position. Now, our simulation with *multiple core convert* is surely not identical to that extreme scenario—yet it comes close, in any case much closer than the purely *random argumentation*. To the degree that the argumentation strategy employed directly targets core positions in order to render them incoherent and is successful in doing so, robustness ceases to be an accurate indicator of long-term stability and proximity to a debate's final consensus. Only if arguments do not always and automatically lead to the falsification of core positions—for example, because they take off from auxiliary, noncore beliefs which might be modified while retaining the core or because they are directed at other proponents—can the robustness of a proponent's core be telling at all. Since arguments which are put forward in line with *multiple core convert* lead much more frequently to an inevitable falsification of a core position than those which are constructed randomly, robustness becomes less useful an indicator of consensus proximity in the *multiple core convert* ensemble.

Even with *multiple core convert*, however, the robustness of a proponent core position does indicate that the corresponding proponent core is close to the partial position the proponent will eventually hold once there remain no more than two nonidentical proponent core positions in the debate. Yet, this result risks to be misleading. It is important to see at which inferential density there are, in the *multiple core convert* ensemble, no more than two nonidentical core positions. For a quarter of the ensemble's debates, this is already the case at $D = 0.19$, and half of the debates host no more than two different proponent cores at $D = 0.27$. So the positive relationship documented in Fig. 9.7a merely demonstrates that the more robust a proponent's core position at $D = 0.15$, the more it agrees with the core position the proponent will adopt at a slightly greater density. If the proponents employ *multiple core convert*, robustness is at best an indicator of short-term durability but not of a partial position's proximity to the full and final consensus.

But how can we explain, finally, the strong correlation between robustness and expected closeness to the final consensus for *robust argumentation*? I suggest that *robust argumentation* combines the best of both worlds: On the one hand, *robust argumentation* generates, like *multiple core convert*, positions with extreme robustness values (see Fig. 9.12). And the greater a core's degree of justification, the higher its ability to resist future argumentation (which represents a potential threat to its coherency). On the other hand, *robust argumentation* is basically a self-centered argumentation rule, like *fortify*, disregarding the opponent positions. As a consequence, *robust argumentation* leads much less frequently to a falsification of proponent core positions than *multiple core convert*. The former strategy thus avoids the problem which makes robustness a poor indicator of consensus proximity in the case of *multiple core convert*.

Part II
How Do We Know? On the Truth-Conduciveness of Controversial Argumentation

Chapter 10
Introduction to Part II

This chapter introduces Part II of our study, which focuses on the veritistic virtues—rather than, as Part I, on the consensual value—of controversial argumentation. Generally, Part II mimics, in terms of its global organization, Part I, and this introduction, in particular, corresponds closely with Chap. 3. We outline, in a first Sect. 10.1, the overall line of reasoning of Part II by explaining how its chapters build on each other. Given this general orientation, we pinpoint, in Sect. 10.2, the different pieces of evidence which back up the main results concerning truth-conduciveness (cf. Sect. 1.5). The corroborating findings are spread all over Part II. So, Sect. 10.2 links the condensed results reported in the general introduction to the specific simulation studies and analyses carried out in Part II.

10.1 Outline of Part II

The debate dynamics studied in Chap. 11 stem from the most basic implementation of the general simulation design: Arguments are introduced randomly, and proponents opt for the closest coherent position. The simple simulations shall serve as a foil to contrast later more sophisticated debate dynamics. Moreover, we investigate, in Chap. 11, the verisimilitude evolutions in the simulated random debates while taking into account (a) different initial conditions (low vs. high initial truthlikeness of proponent positions) and (b) the specific evolution of the space of coherent positions (compact vs. fragmented debates). Controversial argumentation, we find, tends to increase the proponents' verisimilitude without destroying (coincidentally) high initial truthlikeness. In addition, proponents in compact debates, especially those with dominantly false initial beliefs, reach the truth somewhat more rapidly than their counterparts in fragmented debates. Yet, besides studying the truth-conduciveness of *random argumentation* under various conditions, we will scrutinize whether there are indicators that signal reliably the verisimilitude of a given position at low inferential densities. More specifically, we

G. Betz, *Debate Dynamics: How Controversy Improves Our Beliefs*,
Synthese Library 357, DOI 10.1007/978-94-007-4599-5_10,

analyze in which circumstances do (a) a full consensus reached by some proponents and (b) the stability of a proponent's position accurately indicate the proximity to the true position. All in all, the simulations reveal that both factors may serve as useful indicators of truthlikeness.

Relying on background knowledge in the course of an argumentation allows proponents, or the interpreter who reconstructs a debate, to make use of implicit premisses; this reduces the number of explicit premisses per argument and, consequently, increases the inferential density of the dialectical structure (cf. Sect. 2.5). Accordingly, the role of *tacit* background knowledge is indeed studied in Chap. 11, where we investigate the truth dynamics at high inferential densities (which can only be reached with implicit background knowledge). In Chap. 12, in contrast, we establish background knowledge *explicitly* by fixing the truth values of a proportion of the debate's sentences in agreement with the correct position $\mathcal{T}$. We explore how the veritistic dynamics depend on the extent of the background knowledge thus introduced, simulating debates with different levels of fixed background beliefs. In correspondence to Part I, we may observe that constant background knowledge accelerates the convergence towards the truth and explain this finding in terms of the so-called multiplier effect.

Having studied the veritistic dynamics of *random argumentation* (with and without explicit background knowledge) in Chaps. 11 and 12, we drop, in Chap. 13, the assumption that arguments are discovered randomly and suppose that proponents put forward arguments in line with a specific argumentation strategy. In close analogy to our investigation in Part I, we distinguish and study four argumentation rules: *fortify*, *attack*, *convert*, and *undercut*. We simulate debates with two proponents and examine how the truth-conduciveness of controversial argumentation depends on the strategies chosen by the proponents. The simulation experiments reveal that, besides a proponent's initial verisimilitude and the argumentation rule she has adopted, the initial agreement with her opponent plus the opponent's strategy, too, crucially determine the proponent's ability to reach the truth.

Argumentation strategies pursued in two-proponent debates exercise a significant influence on the veritistic dynamics. In Chap. 14, we investigate whether this holds for many-proponent debates as well. In order to do so, we take the two most truth-conducive argumentation rules studied previously—*convert* and *undercut*—and modify them with a view to many-proponent debates. Specifically, the modified *convert* rule (i.e., *t-multiple convert*) instructs proponents to introduce arguments whose premisses are shared by as many opponents as possible; moreover, the strategy prescribes that an opponent position be *rendered incoherent* once a full consensus has emerged. Likewise, the modified *undercut* rule (i.e., *t-multiple undercut*) stipulates that a proponent undercuts and thence renders incoherent, as many opponent positions as possible. As in the case of *t-multiple convert*, the *t-multiple undercut* rule also entails that a full consensus position, reached by all proponents, be rendered incoherent (if possible). We study how these argumentation strategies compare with a purely *random argumentation*, investigated in Chap. 11, while paying particular attention to the veritistic dynamics, to the accuracy of consensus, and to the accuracy of stability as indicators of truthlikeness. Both argumentation strategies—the simulations confirm—substantially enhance the truth approximation in the course of a debate and increase, in general terms, the viability of the veritistic indicators.

In Chap. 15, we loosen the assumption that the proponents in a debate deem all sentences equally important. More specifically, we presume, in analogy to Chap. 8, that a subset of the sentence pool contains the debate's core theses. Proponents are, accordingly, particularly reluctant to modify their convictions regarding these central claims and prefer, rather, to adjust the truth values they assign to the auxiliary sentences outside the debate's core. The introduction of core beliefs allows us to consider the robustness of the proponents' partial positions (i.e., their degree of justification), and we can investigate its bearing on the veritistic dynamics—which is one of the main purposes of Chap. 15. The conservative update mechanism delays, as compared to the simple random debates, the proponents' convergence to the true core position. More importantly, however, we find that (and explain why) a core position's degree of justification yields, even at low densities, a valuable indicator of the position's verisimilitude.

Having studied the veritistic dynamics of random debates with *lexicographic* update mechanism, we will consider, in the final Chap. 16, argumentation strategies that take the distinction between core and auxiliary sentences explicitly into account. More precisely, we modify, firstly, the highly truth-conducive *t-multiple convert* strategy (cf. Chap. 14) with a view to a debate's core sentences and reconsider, secondly, the argumentation strategy which instructs a proponent to maximize the degree of justification of her core position (see Chap. 9). One of our chief interests consists in learning whether a core position's robustness remains an accurate indicator of verisimilitude once proponents employ the sophisticated argumentation strategies. As we find, the argumentation strategies have both a notable effect on the veritistic dynamics and on the reliability of robustness as a veritistic indicator: They enable the proponents to track down the truth more effectively; but with the adjusted *t-multiple convert* strategy, degree of justification ceases to provide a reliable indicator of truthlikeness.

10.2 Main Results and Their Justification

In what follows, we reproduce, in shortened form, the main results regarding truth-conduciveness from Sect. 1.5 and make out the specific simulation experiments, reported in Part II, which support those results.

> T1 (General Results) In toto, controversial argumentation enables proponents to track down the truth. Individual veritistic dynamics vary substantially from debate to debate and are mainly determined by random factors. Still, different argumentative practices give rise to specific mean verisimilitude evolutions and can thence be characterized statistically.

The proponents' truthlikeness increases with ongoing controversy, on average, in nearly every ensemble studied in Part I. This is true no matter whether we regard mean verisimilitude (cf. Figs. 11.1, 12.1, 13.1, 14.1, 15.1a, and 16.1) or the number of fully correct proponent positions (cf. Figs. 11.2, 12.2, 13.2, 14.2, 15.1b, and 16.1). The random variations within an ensemble are illustrated by Fig. 11.4.

T1.1 (Long Run) Proponent positions converge, in the long run, against the truth. Argumentative practices differ, however, significantly with respect to the speed and timing of the verisimilitude increase.

At a density of $D=1$, there is only one dialectically coherent position (cf. Sect. 2.5), and this is the truth, which cannot be rendered incoherent (cf. Sect. 2.6). However, the results below detail that initial conditions and argumentative practices shape, decisively, the veritistic dynamics at low densities.

T1.2 (Epistemic Deterioration) Controversial argumentation may trigger a temporary loss of, instead of a gain in, verisimilitude. Still, verisimilitude evaporates to a much lesser degree than mutual agreement in the course of a debate.

In random debates, controversial argumentation doesn't bring down coincidentally high initial verisimilitude (cf. Figs. 11.1 and 12.1). This contrasts, as discussed and explained in Sect. 11.3, with the corresponding consensus dynamics. Specific argumentation strategies, however, might lead to a temporary loss of high initial verisimilitude (see Sects. 13.2 and 14.2, specifically Figs. 13.4 and 14.1).

T1.3 (Engine of Progress) Criticism is the main driver of epistemic progress. The pace at which proponents approach the truth is largely determined by the frequency at which their positions are rendered incoherent (successfully criticized). Rendering a proponent position incoherent requires, however, that one pinpoints an internal inconsistency pertaining to a subset of the proponent's beliefs, not all of which must, as deductive logic has it, be true. The fact that not all sentences figuring in an alleged inconsistency may be true, whereas, of course, they may all very well be false, amounts to a small but nonetheless influential asymmetry, which assures that, on average, internal critique tends to target more false than correct beliefs and thus prompts a proponent to modify her position to the better.

The import of being criticized is observed in Sects. 13.2, 14.2, and 16.2. We detail the mechanism sketched above in Sect. 11.3 and, specifically, Sect. 13.3 (see also T3.1 below).

T1.4 (Consensual and Veritistic Value) The relationship between consensus- and truth-conduciveness is intricate. A highly truth-conducive practice is necessarily consensus-conducive. Yet, consensus-conduciveness alone does not guarantee truth-conduciveness and can, in fact, prevent proponents from approaching the truth. Argumentative practices which are highly effective in promoting agreement tend to generate spurious consensus.

The detrimental veritistic effect of highly consensus-conducive practices is discussed in Sect. 13.3. Moreover, the simulations in Chaps. 14 and 16 demonstrate that explicit avoidance of spurious consensus increases truth-conduciveness. We refer to spurious consensus, too, when explaining the accuracy of stability as an indicator of truthlikeness (cf. Sects. 11.3 and 14.3).

T1.5 (Space of Coherent Positions) As in the case of consensus-conduciveness, the degree of fragmentation of the space of coherent positions exerts a marked influence on a debate's veritistic dynamics and represents thus a pivotal explanatory variable. As a rule (with several notable exceptions, though), debates with a highly fragmented space of coherent positions display lower verisimilitude increase.

We find, as shown in Figs. 11.3, 12.3, and 14.3, that argumentation tends to be more truth-conducive in compact debates. The notion of the space of coherent positions plays a prominent role in the discussion of Sects. 11.3, 12.3, and 14.3 (see in particular Figs. 11.12 and 14.11).

T2 (Background Knowledge) Background knowledge affects an argumentation's truth-conduciveness in similar ways as its consensus-conduciveness.

We study the effect of fixed background beliefs in Chap. 12.

T2.1 (Multiplier Effect) Constant background knowledge does not simply increase the mean verisimilitude of proponents by a fixed amount but accelerates their approaching the truth, since ever more sentences can be derived from the constant body of background beliefs during a debate.

Section 12.2 reports the acceleration due to background knowledge, which can be explained by the multiplier effect (cf. Sect. 12.3 and Fig. 12.6).

T2.2 (Favorable Fragmentation) With sufficiently many correct background beliefs, the fragmentation of the space of coherent positions turns out to be favorable rather than detrimental to an argumentation's truth-conduciveness.

The inverted effect of fragmentation (see Figs. 12.3 and 12.4) is explained, in Sect. 12.3, by means of the fishing-net and the flooded-village metaphor.

T3 (Argumentation Strategies) The veritistic value of an argumentative practice does not correspond, one to one, with its consensual value. A proponent's ability to track down the truth is determined by her own argumentation strategy as much as by her opponents' ones.

The interrelation between consensus- and truth-conduciveness of argumentation strategies is, first, studied with a view to dualistic debates in Chap. 13 and elaborated in Chaps. 14 and 16.

T3.1 (Veritistic Value of Critique) As the advancement towards the truth is primarily driven by criticism, proponents whose positions are frequently rendered incoherent exhibit a comparatively rapid verisimilitude increase. In consequence, it is the argumentation strategy employed by one's opponent, and this opponent's ability to advance critical arguments, which controls the pace at which one acquires more and more true beliefs.

Section 13.2 finds that proponents whose opponents argue in an aggressive and opponent-sensitive way (*undercut* rule) display the strongest verisimilitude rise (cf., in particular, Fig. 13.3). Opponents, in contrast, who don't address a proponent's position at all, arguing in a self-centered way, don't allow the proponent to improve her position. See, for a discussion, Sect. 13.3 (but compare also Sects. 14.3 and 16.3).

T3.2 (Veritistic Value of Plurality) Outstanding consensus-conduciveness and the inability to question (and give up) a reached consensus contribute to an

(continued)

(continued)
argumentative practice's consensual value but tend to curtail its veritistic one. This is strikingly revealed by our simulations, where proponents who implement the *convert* rule fare poorly in terms of verisimilitude. Now, high initial disagreement and the employment of agreement-reducing strategies, side by side with consensus-conducive ones, can help to avoid the emergence and persistence of a spurious consensus and enable proponents to continue questioning their beliefs. Plurality, we find, is an instrumental epistemic virtue, and argumentative practices which explicitly cultivate it (in an, otherwise, extremely consensus-conducive climate) foster a debate's overall truth-conduciveness.

The poor veritistic performance of the *convert* rule is revealed in Sect. 13.2. This section shows, moreover, that the agreement-reducing *attack* strategy and high initial disagreement (i.e., plurality) may help proponents to track down the truth (cf. Fig. 13.6). Further discussions, such as in Sect. 14.3, confirm the general relevance of plurality and the perils of spurious consensus.

T3.3 (Consensus First) Aggressive and opponent-sensitive argumentation (i.e., the *undercut* strategy) represents the most truth-conducive practice in dualistic debates. This is, however, not the case if multiple proponents engage in a controversy. Instead of fervently criticizing the various proponent positions simultaneously, it is more efficient to generate a consensus, possibly a spurious one, in a first step and to criticize the consensus position (by way of self-critique) in a second step. This more conciliatory strategy, it turns out, is, in sum, more truth-conducive than an immediate criticism of the diverse proponent positions. A specific version of the *convert* rule has, consequently, a role to play in truth-seeking controversies as well.

The superior truth-conduciveness of the *t-multiple convert* strategy, as compared to *t-multiple undercut*, is revealed and explained in Chap. 14.

T4 (Veritistic Indicators) We may identify three veritistic indicators, which signal the truthlikeness of a proponent's position at low densities: consensus, stability, and degree of justification. Remarkably, these indicators suggest a novel, "dialectic" foundation of the two major methodologies which have been developed in philosophy of science, that is, falsificationism and verificationism.

Because—on average and irrespective of the argumentative practice employed—proponent positions approach the truth only in a relatively advanced phase of a debate and since, in addition, real debates (for lack of new arguments) often don't attain these advanced phases, it becomes a decisive question whether there are reliable methods for gauging the verisimilitude of proponent positions in an early stage of a debate. Sections 11.2 and 14.2 explore whether stability (of proponent positions) and consensus may serve as indicators of truthlikeness; Chaps. 15 and 16 focus on robustness (i.e., degree of justification) as a veritistic indicator.

T4.1 (Consensus) Consensus, for being possibly spurious, may obviously be a misleading indicator of truth. Still, a consensus which is reached not simply by two but by at least five or six (independently arguing) proponents is typically a very good indicator of truth. In general, the greater the size of a consensus (in terms of proponents), the higher its expected verisimilitude. If the proponents who reach the consensus display substantial initial disagreement, the reliability tends to improve even further. The accuracy of consensus as an indicator of truth depends, moreover, on the specific argumentation strategy employed by the proponents. The more consensus-conducive the argumentative practice, the less reliable the indicator. In a highly critical controversy (proponents follow the *undercut* rule), however, even a two-proponent consensus represents a highly accurate indicator of truth, especially at an early stage of the debate.

The accuracy of consensus as a veritistic indicator is a function of consensus size (as shown in Figs. 11.6 and 14.4) and initial agreement among consensus members (cf. Figs. 11.7 and 14.5). These observations stress, in addition, the impact of the argumentation strategy on the indicator's reliability (see also the discussions in Sects. 11.3 and 14.3).

T4.2 (Stability) While a proponent position's stability indicates, in general, truthlikeness in a reliable way, its accuracy depends on the argumentation strategies pursued by the debate's proponents. Specifically, the more critical the argumentation, the more accurate the indicator. With proponents who implement the *undercut* strategy, stability becomes in fact an extremely reliable indicator of truth. This allows us to make sense and to justify core tenets of a refined falsificationist methodology.

The stability of a proponent position can be measured in different ways—as agreement of the position with the proponent's initial position or as relative frequency at which the proponent had to modify her previously held positions (cf. Sect. 11.2). No matter how one gauges stability, however, it yields a telling indicator of a position's verisimilitude at an early stage of a debate (see Figs. 11.10, 11.11 and 14.7–14.10), whose reliability improves, moreover, as an argumentation becomes more critical (cf. Sects. 14.2 and 14.3).

T4.3 (Degree of Justification) The verisimilitude of a proponent's core position is, at an early stage of a debate, correlated with its degree of justification. Degrees of justification thus signal proximity to the truth. The correlation between degree of justification and verisimilitude is particularly strong if arguments are discovered randomly or introduced by proponents with a view to maximizing their positions' robustness.

Figures 15.3 and 16.4b demonstrate, for *random* and, respectively, *robust argumentation*, the significant correlation between the verisimilitude of a core position and its degree of justification. We explain, in Sect. 15.3, why degree of justification reliably indicates truthlikeness in random debates.

T4.4 (Methodological Trade-Off) In random debates, both stability and degree of justification may serve as accurate veritistic indicators. Yet, if one attempts to sharpen the accuracy of stability by stipulating that proponents argue in a highly critical way, the reliability of degree of justification as an indicator of truth is completely lost. There is a certain trade-off between the two indicators, since the accuracy of the indicators hinges sensitively on the argumentation strategies pursued by the proponents.

We observe, in Sect. 16.2, that the degree of justification ceases to be a useful indicator of truth as soon as proponents start to argue in a more critical, purposeful way, employing *t-multiple core convert* (compare, specifically, Figs. 16.4a, b). Section 16.3 discusses and explains this observation.

Chapter 11
The Veritistic Dynamics of Simple Random Debates

The debate dynamics studied in this chapter stem from the most basic implementation of the general simulation design: Arguments are introduced randomly and proponents opt for the closest coherent position. The simple simulations shall serve as a foil with which we may contrast later, more sophisticated debate dynamics. Moreover, we investigate, in this chapter, the verisimilitude evolutions in the simulated random debates, while taking into account (a) different initial conditions (low vs. high initial truthlikeness of proponent positions) and (b) the specific evolution of the space of coherent positions (compact vs. fragmented debates). Yet, in addition to studying the truth-conduciveness of *random argumentation* under various conditions, we will scrutinize whether there are indicators that signal reliably the verisimilitude of a given position at an early stage of the debate. More specifically, we analyze in which circumstances do (a) full consensus reached by some proponents and (b) stability of a proponent's position accurately indicate the proximity to the true position.

11.1 Setup

Debate simulations are initialized by determining the proponents' initial positions (cf. Footnote 10 on page 49) and marking a randomly chosen assignment of truth values as the truth $\mathscr{T}$.[1]

Argumentation Mechanism: Arguments are constructed randomly. At each step, a single argument a is drawn randomly from the set of all arguments which, if

[1]See also Sect. 2.6. Technically, the truth is identified with the Boolean vector $\langle$True, ..., True$\rangle$ in the simulation code. This is justified insofar as, once a true position is chosen randomly, proposition variables can be redefined ($p'_i := \neg p_i / p_i$) so that the true position becomes $\langle$True, ..., True$\rangle$ on the relabeled sentences.

G. Betz, *Debate Dynamics: How Controversy Improves Our Beliefs*, Synthese Library 357, DOI 10.1007/978-94-007-4599-5_11,

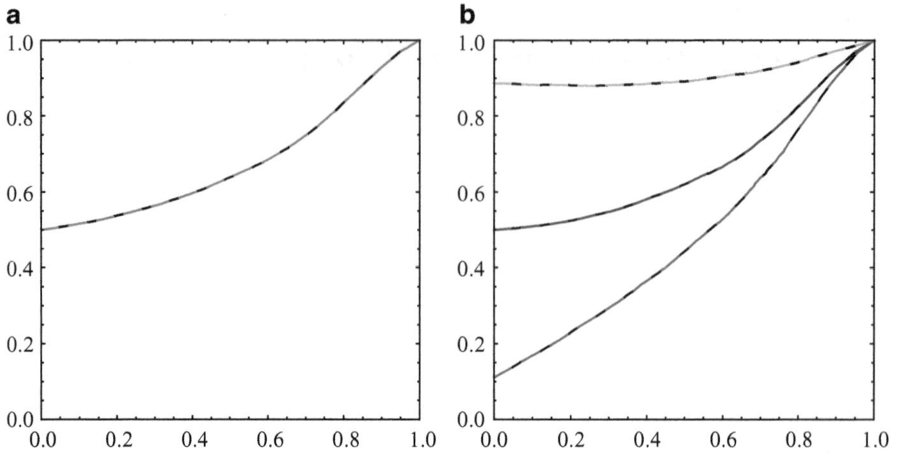

Fig. 11.1 Ensemble-wide mean verisimilitude of proponent positions as a function of inferential density. The mean verisimilitude at a given density averages, at that very density, the proponents' normalized agreement with the true position $\mathcal{T}$. The curve in plot (**a**) takes account of all proponent positions in the ensemble's debates. The three curves in the *right-hand plot*, however, average but over proponent positions with high (0.8–1, *top*), medium (0.4–0.6, *middle*), and low (0–0.2, *bottom*) initial verisimilitude

added to the dialectical structure, leaves the true position, $\mathcal{T}$, dialectically coherent, and argument a is introduced into the debate. We call this argumentation mechanism *t-random argumentation*.

Discovery Mechanism: There is no background knowledge.

Update Mechanism: *Closest coherent* (cf. Sect. 4.1).

Each debate contains six proponents. It terminates once all proponents hold the true position $\mathcal{T}$. One thousand debate simulations in accordance with these specifications yield the ensemble we study in this chapter.

11.2 Results

11.2.1 Truth's Attraction: How Rapidly Does the Proponents' Verisimilitude Increase?

Do proponents gradually approach the truth during a debate? And how rapidly does their verisimilitude increase? Fig. 11.1a provides an answer to this question. It displays the mean verisimilitude (averaged over all proponents in all debates) as a function of inferential density. Verisimilitude takes off at an initial value of 0.5—in line with the fact that initial proponent positions are sampled independently of the truth. It rises gradually to 0.65 at a density of $D = 0.5$ and reaches, accelerating

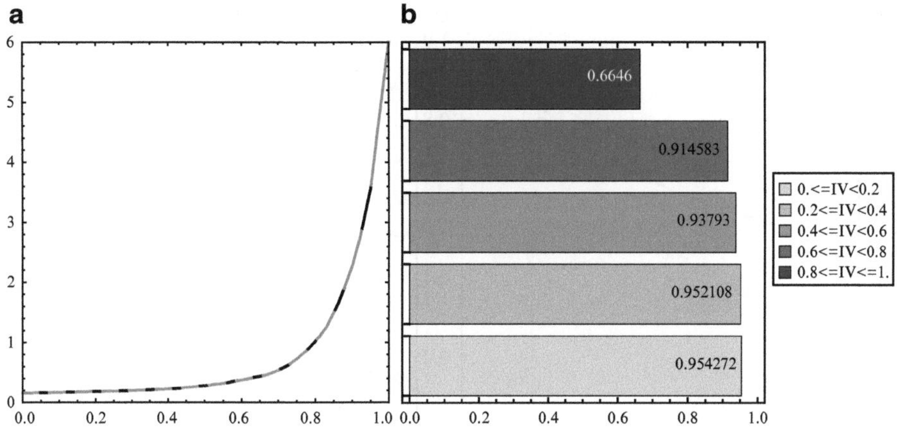

Fig. 11.2 Plot (**a**): Ensemble-wide mean number of entirely true proponent positions per debate as a function of inferential density. Plot (**b**): Ensemble-wide mean densities at which proponent positions that possess a certain initial verisimilitude collapse onto the true position

its incline, 100% agreement with the truth at a density of $D = 1$, that is, at a situation where the true position is the only remaining coherent position in the debate. Like in the case of mean agreement, the major increase in mean verisimilitude occurs at densities greater than 0.5.

Figure 11.1a averages the verisimilitude values of all proponent positions, irrespective of their initial verisimilitude. Yet, the initial agreement with the true position might exert a notable influence on the further verisimilitude of a proponent's position. That is why, in Fig. 11.1b, we distinguish proponent positions with high (0.8–1), medium (0.4–0.6), and low (0–0.2) initial verisimilitude and consider their respective verisimilitude evolutions. Proponents with medium initial verisimilitude (middle curve) approach the truth only slowly at low densities before, finally, advancing towards the truth with ever bigger steps. Proponent positions that encompass, initially, very few correct individual beliefs, however, get significantly closer to the truth at low densities, as well. Thus, at $D = 0.5$, their verisimilitude has increased from ca. 0.1 to approx. 0.45. Finally, positions with relatively high initial verisimilitude retain that level at low densities and approach the truth gradually at $D > 0.5$. And here lies the main qualitative difference to the consensus dynamics of random debates. Unlike coincidental mutual agreement, coincidental verisimilitude is not systematically and significantly destroyed by controversial (random) argumentation.

Figure 11.1 displays the veritistic dynamics in terms of mean verisimilitude of proponent positions. Counting the number of proponents which have acquired the true position (verisimilitude = 1) yields an alternative perspective on the debates' dynamics. As Fig. 11.2a shows, the number of fully true proponent positions per debate increases only very slowly. No earlier than at a density of $D = 0.8$ has at least one proponent per debate, on average, found the truth. Clearly, all proponents will eventually adopt the truth when no other position is coherent, which results

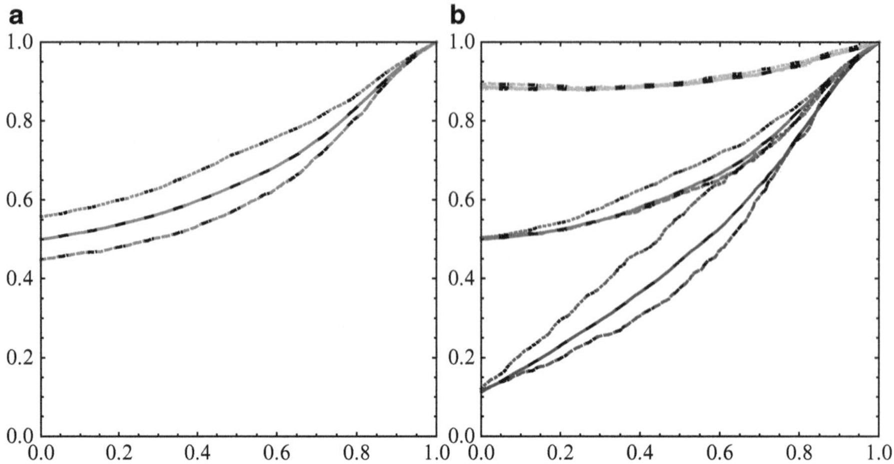

Fig. 11.3 Ensemble-wide mean verisimilitudes as a function of inferential density. *Solid curves* are calculated with respect to all debates in the ensemble and thence correspond to Fig. 11.1a, b. *Dashed* and *dotted curves*, however, average over a subset of the ensemble's debates only. More specifically, *dotted curves* (*dashed curves*) depict the mean evolutions averaged over the 100 most compact debates (most fragmented debates), as measured by aggregated NCC

in the steep incline at very high densities. Figure 11.2b provides some more detailed information by distinguishing different initial verisimilitude values. It plots, more precisely, the inferential densities at which a proponent position with the corresponding initial verisimilitude typically collapses onto the true position. Even proponents who are initially very close to the truth don't adopt a fully correct position, on average, before a density of $D = 0.66$ has been reached. The corresponding collapse-to-truth densities for proponents with lower initial verisimilitude are significantly greater and lie well above 0.9. In sum, the prospect of acquiring a fully true position by *t-random argumentation* appears to be rather desolate.

We found, in Chap. 4, that the way arguments carve the space of coherent positions influences the debate's consensus dynamics crucially. In particular, debates with a relatively compact SCP displayed more constant and rapid rapprochement than debates whose SCP has been fragmented. The metaphors of the fishing net which is pulled and the village which is flooded served as helpful analogies to understand these dynamics. So, it is natural to ask how the fragmentation of the SCP affects the veritistic dynamics of debates. Figure 11.3 plots the evolutions of mean verisimilitude as calculated with respect to all debates (solid), with respect to relatively compact debates (dotted), and with respect to highly fragmented debates (dashed). Concerning the mean verisimilitude evolution of proponent positions with arbitrary initial verisimilitude, shown in panel (a), the key difference between debates with compact and fragmented SCP roots in the corresponding initial values. The compact debates possess, initially, a somewhat above-average verisimilitude (0.55), fragmented debates a lower one (0.45).

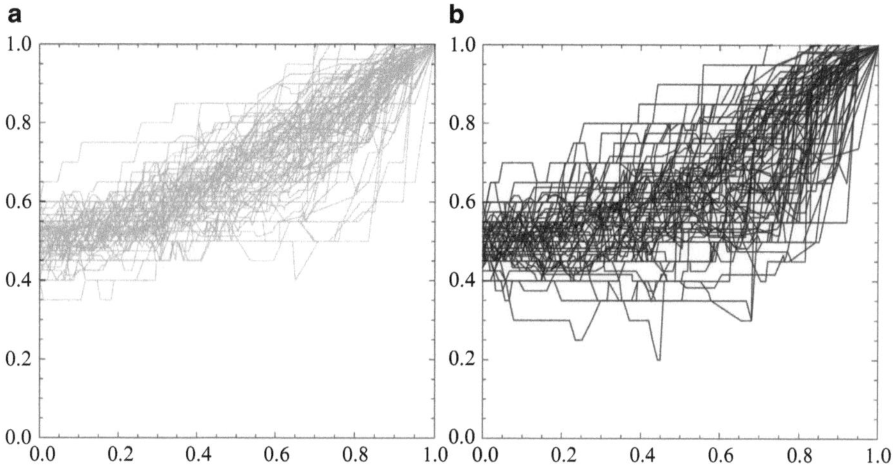

Fig. 11.4 Debate-wide mean verisimilitude as a function of inferential density. The *left-hand plot* displays verisimilitude evolutions of compact debates (100 debates with highest aggregated NCC), the *right-hand plot* those of fragmented debates (100 debates with lowest aggregated NCC). Each curve represents a debate-wide mean verisimilitude, averaged over all proponent positions in the respective debate whose initial verisimilitude is greater than 0.4 and less than 0.6. As a consequence, the *dotted* (*dashed*) *middle curve* in Fig. 11.3b represents the mean of the evolutions in the *left-hand* (*right-hand*) *panel* above

In addition, the verisimilitude increase at densities below 0.5 appears to be slightly greater in compact as compared to fragmented debates, namely, 17 versus 12 percentage points. Yet, it is not clear whether this amounts to a significant difference at all. We see, however, some more distinct differences if we consider proponent positions with specific initial verisimilitude values, as shown in Fig. 11.3b. While the verisimilitude of proponent positions which are originally close to the truth seems to evolve quite similarly in compact and fragmented debates (top curves), the differences between these two types of debates become ever more pronounced as one considers proponent positions with lower initial verisimilitude. Thus, positions with medium initial verisimilitude (middle curves) seem to approach the truth significantly more steadily in compact than in fragmented debates. This difference is even more striking for positions with very low initial verisimilitude (bottom curves): In compact debates, the verisimilitude of such positions has increased by roughly 45% at a density of $D = 0.5$. In fragmented debates, however, the corresponding increase of verisimilitude amounts to merely 25 percentage points.

Figure 11.4 provides an even more detailed picture of how proponent positions with medium initial verisimilitude (0.4–0.6) approach the truth in compact debates on the one side (a) and in fragmented debates on the other side (b). Obviously, the verisimilitude of proponent positions increases much more steadily and continuously in compact debates, while proponents tend to approach the truth stepwise, with more or less greater leaps, in debates whose SCP is fragmented.

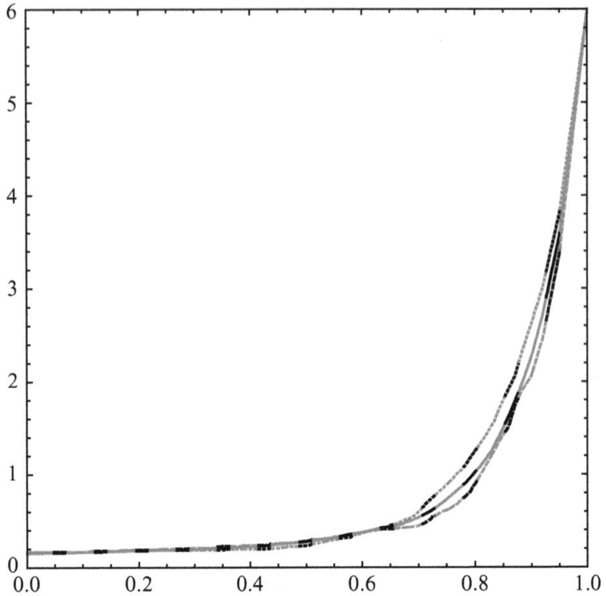

Fig. 11.5 Ensemble-wide mean number of entirely true proponent positions per debate as a function of inferential density. The *solid curve* is calculated with respect to all debates in the ensemble and thence corresponds to Fig. 11.2a. *Dashed* and *dotted curves*, however, average over the most fragmented and, respectively, most compact debates only

Finally, regarding the number of proponents who have found the truth, debates with compact and fragmented SCP are virtually indistinguishable, as shown by Fig. 11.5.

11.2.2 The Verisimilitude of Consensus Positions: Is Mutual Agreement a Good Indicator of Having Reached the Truth?

In real debates, agreement is, from time to time, considered an indicator of truth: If the opponents have, after a controversial argumentation, eventually settled on a consensus, the consensus position is likely to be true, or so it seems. Our simulations allow us to scrutinize this claim. Does mutual agreement really signal that proponents have settled on the truth, or under which circumstances does it do so? It seems plausible that a consensus is ever more telling, the more proponents have joined it. Figure 11.6 verifies this hypothesis. Its left-hand plot displays the relative frequency at which a consensus position is identical with the true position as a function of the consensus' size. Obviously, the more proponents have come to agree, the greater the likelihood that they have found the truth. Surprisingly, even a consensus among five

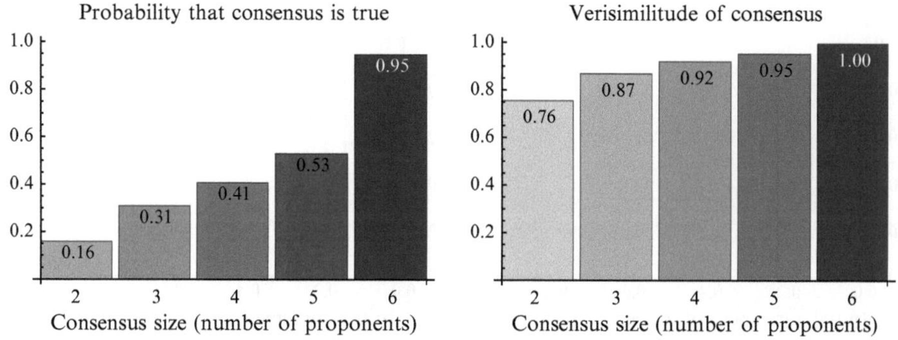

Fig. 11.6 Ensemble-wide relative frequency at which a consensus represents the true position as a function of the number of proponents who have come to agree (*left-hand plot*) and ensemble-wide mean verisimilitude of a consensus position as a function of the number of proponents who have come to agree (*right-hand plot*)

proponents is in merely half of the instances identical with the truth. A consensus among six proponents, however, is much more telling: it is very likely (95%) to be true. The left-hand plot simply counts the consensus positions which are identical with the true position, ignoring different degrees of verisimilitude altogether. Not so the right-hand plot of Fig. 11.6, which shows the mean verisimilitude of a consensus with a given size and therefore provides additional information. Consensus positions of no matter which size exhibit, on average, verisimilitudes considerably greater than 0.5. Moreover, consensus verisimilitude depends positively on consensus size. So, for example, a 2-proponent-consensus assigns a correct truth value to 76% of the debate's sentences. And although only half of the consensus positions among five proponents represent the truth, as noted above, the mean verisimilitude of these consensus positions is almost 1 (precisely 0.95). This signifies that a consensus among five proponents is typically almost true.

Consensus size is only one among many factors which potentially influence how accurately mutual agreement indicates truth. In the following, we consider two further factors—namely, initial agreement between the proponents who belong to the consensus and inferential density at which the consensus emerges—restricting the analysis, in a first step, to 2-proponent-consensus.

Figure 11.7 displays the influence of initial agreement among consensus members. The left-hand plot shows that the initial agreement has no significant effect on whether a consensus reached by two proponents is identical with the truth or not. But, as the right-hand plot reveals, initial agreement does have at least a weak effect on how close such a consensus position is to the truth—just as we would expect. The greater the initial agreement between the consensus members, the lower the verisimilitude of the consensus itself. That is, if, coincidentally, consensus members have agreed initially to a large extent, the consensus tends to be spurious and doesn't signal high verisimilitude. In the extreme case where the two proponents exhibit full initial agreement, the consensus has emerged, by pure chance, at the very first step of the debate and possesses, on average, and unsurprisingly, a verisimilitude of 0.5.

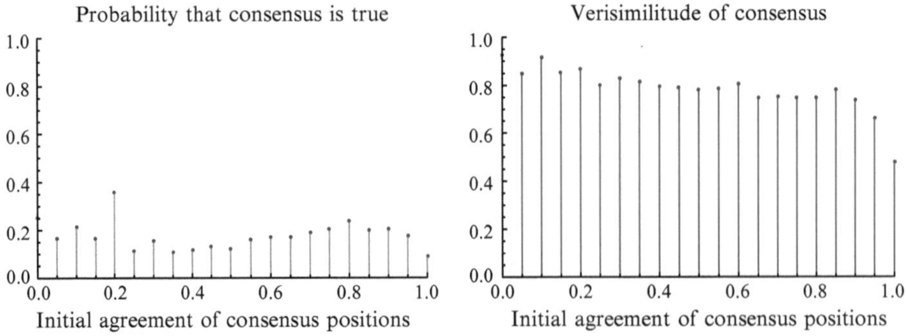

Fig. 11.7 Ensemble-wide relative frequency at which a 2-proponent-consensus represents the true position (*left-hand chart*) and ensemble-wide mean verisimilitude of a 2-proponent-consensus position (*right-hand chart*), both plotted as a function of initial agreement between the two proponents who reach the consensus

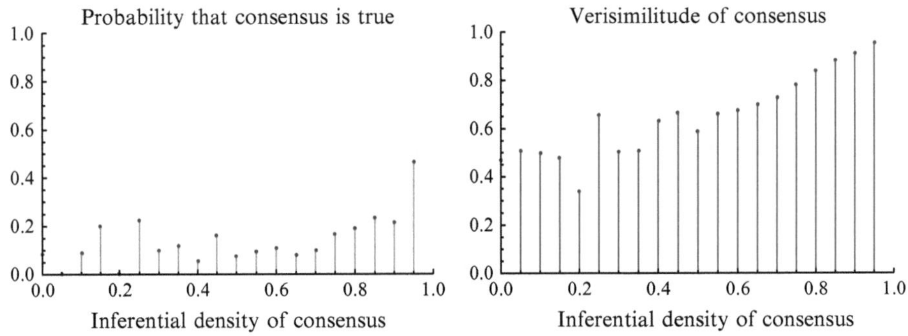

Fig. 11.8 Ensemble-wide relative frequency at which a 2-proponent-consensus represents the true position (*left-hand chart*) and ensemble-wide mean verisimilitude of a 2-proponent-consensus position (*right-hand chart*), both plotted as a function of the inferential density at which the corresponding consensus emerges

The accuracy by which consensus indicates truth depends positively on consensus size and negatively on initial agreement between consensus members. A third factor, whose impact we will gauge in the following, is the inferential density at which a consensus is reached. Figure 11.8 demonstrates that the inferential density at which a 2-proponent-consensus emerges does indeed affect the consensus' accuracy in terms of an indicator of truth. More precisely, the likelihood that a consensus reached between two proponents represents the truth depends, at least for densities greater than 0.5, positively on the consensus' inferential density (left-hand plot). The influence of a consensus' density on its (expected) verisimilitude is even more pronounced (right-hand plot). If two proponents come to agree at low densities ($D < 0.4$), their consensus position exhibits, on average, a verisimilitude of 0.5, that

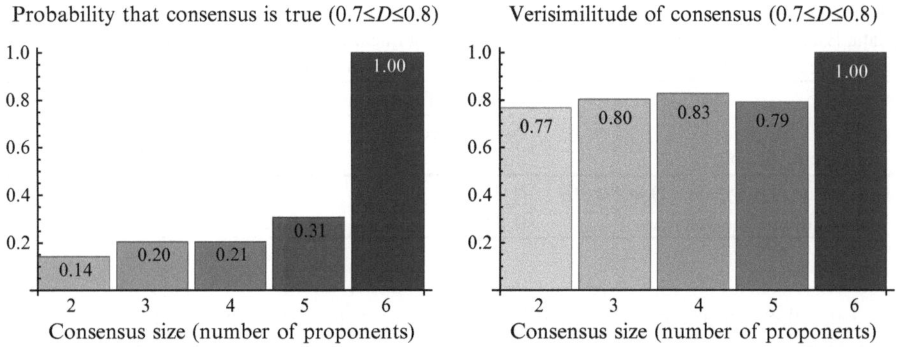

Fig. 11.9 Ensemble-wide relative frequency at which a consensus that emerges at an inferential density of 0.7–0.8 represents the true position (*left-hand plot*) and ensemble-wide mean verisimilitude of consensus positions which emerge in the corresponding density interval (*right-hand plot*), both plotted as a function of the number of proponents who have come to agree

is, no more than a random position. With increasing density, the verisimilitude of a 2-proponent-consensus gradually climbs to 1, mimicking, approximately, the mean verisimilitude evolution plotted in Fig. 11.1a.

The findings reported so far might suggest that the primary factor which determines the accuracy of consensus as an indicator of truth is neither consensus size nor initial agreement but rather inferential density. For a broad consensus tends to occur at higher densities, and the smaller the initial agreement between some proponents, the longer it takes to reach a consensus. So, is the positive correlation between, say, consensus verisimilitude and consensus size merely the result of the facts that (1) proponents exhibit, at high densities, higher verisimilitude and (2) broad consensus tends to materialize at high densities only? Or does the fact that proponents have reached an agreement provide an additional, independent reason for considering the consensus position as correct?

In order to answer this question, we consider, in Fig. 11.9, but consensus positions which are reached in the density interval $[0.7;0.8]$. As the left-hand plot shows, there is still a positive relationship between consensus size and the probability that the respective consensus represents the truth—even if we disregard, as a concession to the small sample size, the consensus among six proponents. However, the positive relationship between consensus size and verisimilitude, which we had previously identified, is much more difficult to discern once we keep the density constant, as the right-hand plot illustrates. A greater number of proponents doesn't automatically imply that the consensus reached is closer to the truth. In sum, Fig. 11.9 suggests that a consensus' size provides, relative to the inferential density at which the consensus emerges, more information about the probability of the consensus position being true than about its verisimilitude. A linear regression analysis confirms this result. A fitted linear model which explains that a given consensus is true as a function of (a) normalized consensus size (six proponents $\sim$1), (b) initial agreement among consensus members, and (c) inferential density at which the consensus emerges

Table 11.1 Parameters of independent variables in a linear model that is fitted to the ensemble data. Each row displays the values corresponding to a linear model which explains that a consensus represents the truth, respectively, its verisimilitude, in terms of the three independent variables

	Weights of independent variables		
Dependent variable	Normalized consensus size	Initial agreement	Inferential density
Consensus is fully true (1) or not (0)	0.96	−0.25	0.41
Verisimilitude of consensus	0.10	−0.05	0.58

assigns the parameters displayed in Table 11.1 to these three factors.[2] Hence, whether some consensus represents the truth can primarily be explained by its size; consensus size is, more precisely, twice as important as the inferential density at which the consensus emerges and, in absolute terms, roughly four times as influential as the initial agreement among consensus members (which exerts a negative influence). A consensus' verisimilitude, however, seems to be crucially determined by its inferential density, as the parameters of the corresponding linear model suggest (cf. Table 11.1). Inferential density is 6 times more important an explanatory factor than consensus size and even 12 times more important than initial agreement between consensus members.

11.2.3 The Verisimilitude of Stable Positions: Are Proponent Positions Which Remain Relatively Stable Closer to the Truth?

The results reported in the previous sections possess the following common feature: They allow us to tell which position is true, or at least close to the truth, only once relatively high inferential densities have been reached. For only at high densities have proponents significantly approached the truth, and only at high densities do many proponents agree on a shared consensus position. From a methodological point of view, it would, however, be highly desirable if we could gauge the verisimilitude of some positions—or estimate, with some degree of accuracy, which position represents the truth—at an early stage of the debate, that is, at low inferential densities. Since this would obviously allow us to learn from controversial debates which, lacking sufficiently broad (implicit) background knowledge, don't reach high inferential densities.

By suggesting that the more severe tests a theory has successfully passed without being falsified, the closer the theory is to the truth, Karl Popper has actually hypothesized a method for doing so (Popper 1963, p. 333). In a nutshell, if a theory isn't falsified in a critical argumentation, it tends to be close to the truth. Or, to

[2]Note that all independent variables vary between 0 and 1, which allows to interpret the parameters as commensurate proxies of the variables' influence.

put it differently, the more stable a proponent position, the higher its verisimilitude. That's the hypothesis we are going to examine in this section. While doing so, we quantify a proponent position's stability in alternative ways. A first, straightforward measure of stability is the relative frequency at which the proponent had to modify her beliefs because her previous position had been rendered incoherent (falsified). Alternatively, we express a proponent position's stability by its agreement with the proponent's initial position: The farther a proponent has retreated from her initial position, the less stable is her current position.

The fan charts in Fig. 11.10 depict the relationship between a position's stability (understood as frequency of previous falsifications) on the one hand and its verisimilitude on the other hand at different levels of inferential density. By marking quantiles, the different shadings visualize the verisimilitude distribution of core positions that possess a similar stability value (x-axis). In the absence of any arguments ($D = 0.0$), no proponent position has obviously been rendered incoherent, and all positions possess maximum stability, 1, and are distributed over the verisimilitude interval in line with the initial sampling (cf. Footnote 10 on page 49), yielding a high-verisimilitude proportion of $q = 0.16$ (see below). As arguments are introduced into the debate, some proponents have to modify their positions, and others don't. As the plots demonstrate, a positive relationship emerges even at low inferential densities (maybe not yet at $D = 0.05$ but clearly at $D \geq 0.1$): The less stable a proponent position, that is, the more frequently it had to be modified, the smaller its expected verisimilitude and the less likely it is close to the truth. This positive relationship seems to hold up to a density of $D = 0.6$, before it breaks down at higher inferential densities. What's more, high stability (a proponent position is falsified, on average, by less than 1 out of 10 arguments) becomes ever more accurate an indicator of high verisimilitude as inferential density increases. At the beginning, 16% of the highly stable positions (i.e, in the absence of arguments, 16% of all positions) possess a verisimilitude greater than 0.8. This proportion climbs to roughly a quarter for densities between 0.05 and 0.2. And at $D = 0.5$, almost half of the positions with high stability deviate by less than 20% from the truth.

Even more telling, and interesting, an indicator of verisimilitude is, however, a proponent position's stability as measured by its agreement with the proponent's initial position. The plots in Fig. 11.11 visualize the way verisimilitude depends on how far a proponent had to retreat from her original position. Once again, we observe a positive relationship for low densities greater than zero: The greater the stability, the higher the expected verisimilitude. But note that, at low densities, the stability is never less than 0.5. As we move to densities greater than 0.2, the relationship changes in an unexpected way: The positive relationship between stability and expected verisimilitude prevails for stability values greater than 0.5. It does, however, not hold for stabilities below 0.5. On the contrary, very low stabilities seem to be an accurate indicator of high verisimilitude. Thus, if a proponent has altered 80% of her initial beliefs at $D = 0.6$, adopting a position with stability value 0.2, her position is very likely very close to the truth. So what we have, in sum, is a nonlinear relationship between stability and verisimilitude: Expected verisimilitude of extremely stable and extremely unstable positions is high; positions with medium

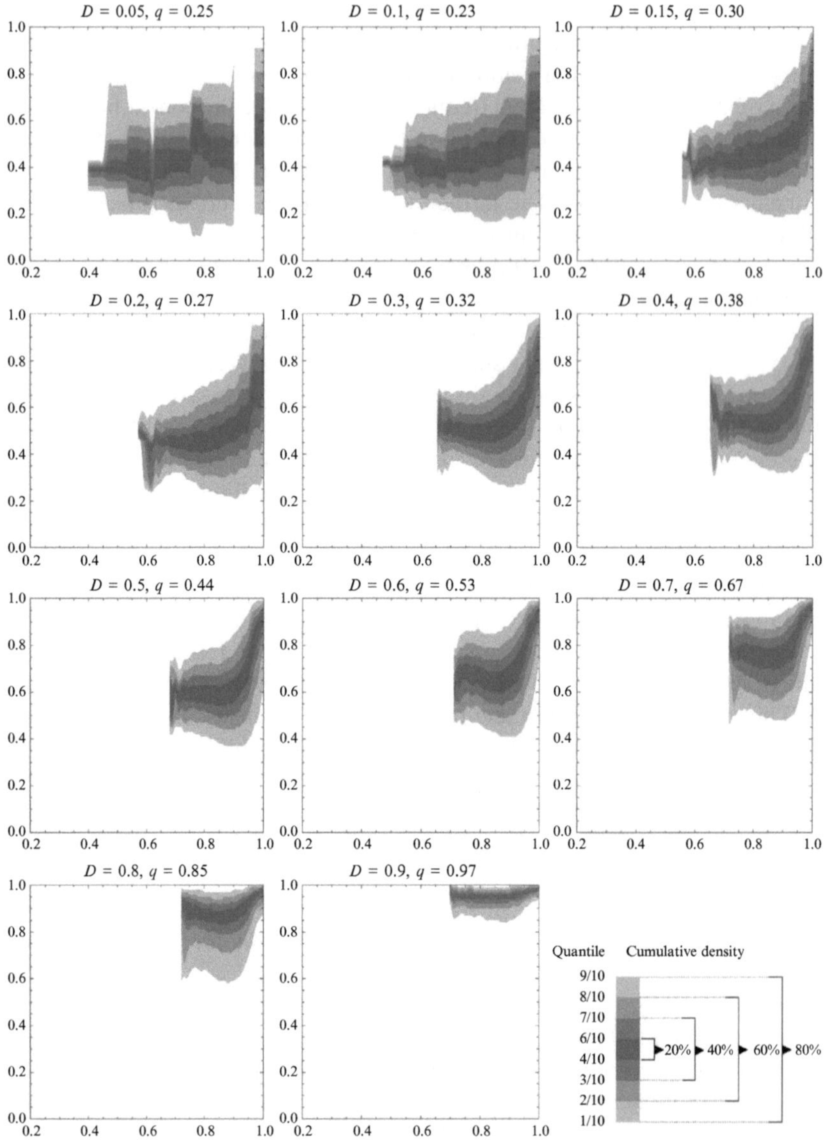

Fig. 11.10 Verisimilitude of proponent positions, at a certain density D, as a function of their stability. Stability is approximated by how frequently previous positions of the corresponding proponent have been rendered incoherent, precisely by 1 minus the corresponding relative frequency. The shadings in the fan chart indicate the different quantiles as specified in the legend. The quantiles are calculated as follows: For each stability value (x-axis), a smooth probability density function (PDF) is fitted to the discrete relative frequencies of different verisimilitude values. This interpolated PDF is then used to derive the quantiles. The values on *top* of each diagram indicate the inferential density (D) as well as the proportion of very stable positions whose verisimilitude is greater than 80%, that is, the discrete relative frequency of positions with stability greater than 0.9 and verisimilitude greater than 0.8 relative to all positions with stability greater than 0.9

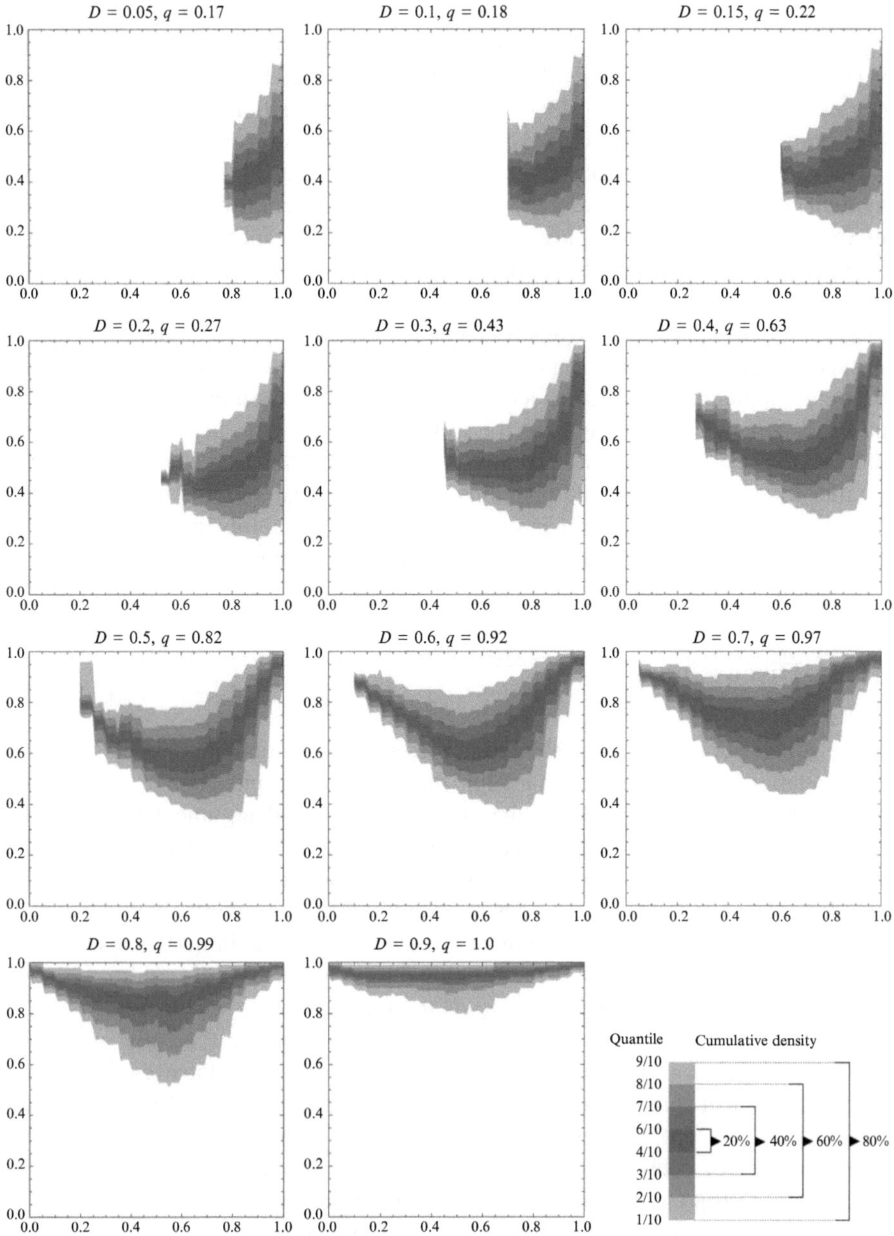

Fig. 11.11 Verisimilitude of proponent positions, at a certain density D, as a function of their stability. The fan charts are constructed as in Fig. 11.10, except that a proponent position's stability is measured by its agreement with the proponent's initial position

stability, however, display a comparatively low mean verisimilitude. Focussing on highly stable positions, we find that agreement with the initial position predicts verisimilitude more accurately than stability as measured by falsification frequency (for $D \geq 0.2$). So, at $D = 0.3$, 43% of all highly stable positions are close to the truth (verisimilitude greater than 0.8). At $D = 0.6$ this proportion exceeds even 90%.

In sum, the findings presented in Figs. 11.10 and 11.11 validate Popper's hypothesis that the stability of a proponent position in a controversial debate reveals its verisimilitude and that highly stable positions are ever more likely to be close to the truth as the critical debate proceeds and proponent positions have been submitted to ever new challenging arguments.

As a caveat, I'd finally like to stress that the precise numerical findings presented in this subsection crucially depend on how initial positions are determined. This becomes apparent if we consider an extreme case. Assume that, initially, no positions whatsoever are close to the truth. In such a case, high stability cannot indicate the truth because there simply is no highly stable, true position at all. The smaller the proportion of true positions in the initial sample, the more difficult it becomes to differentiate between (1) positions that are highly stable because they are true and (2) positions which are highly stable by coincidence.

11.3 Discussion

Some, albeit not all, results presented in the previous section are in need of further interpretation. Thus, the fact that high initial verisimilitude, unlike coincidental agreement, doesn't evaporate in the course of a critical argumentation calls for an explanation. Moreover, we have to discuss why verisimilitude evolutions are influenced by the shape of the SCP. Finally, the finding that not only extremely stable but also extremely unstable positions are likely to be true requires an explanation.

As we've stressed above, proponent positions which are, initially, close to truth retain, on average, their high verisimilitude during a debate. This contrasts starkly with the consensus dynamics of *random argumentation*, where coincidentally high agreement among two proponents in the initial phase typically evaporates as the controversial argumentation unfolds. In Chap. 4, we have explained this decrease of mean agreement by what we called the random walk effect: As closely related positions are rendered incoherent by new arguments, they are modified independently of each other. Because of the relatively great room of maneuver for adjusting proponent positions in the initial phase, two proponents are likely to modify different sentences in order to readjust their positions, thereby destroying their coincidental, high agreement. Now, this mechanism doesn't seem to apply with respect to verisimilitude. But for what reason? As a first thing to recall, verisimilitude does not measure the agreement between two proponent positions but between a proponent position on the one hand and the true position on the other hand. I see two reasons why a position which is initially close to the truth is, on average, not pushed away from the truth by the ongoing argumentation.

Firstly, there exists a lock-in effect as regards verisimilitude. Once a proponent position is identical with the truth, it cannot be rendered incoherent by any argument whatsoever and remains, as a consequence, unchanged. Proponents who have—even by chance—found the truth stick to it. This is obviously untrue for a consensus among proponents, which may very well dissolve. So, consider all proponents with a high initial verisimilitude. Some of these will hold the true position right from the start. Others will coincidentally hit the truth at an early stage of the debate as they are compelled to readjust their position. These proponents will display a constant verisimilitude equal to 1 in the remaining debate and therefore contribute to a relatively high mean verisimilitude of proponent positions which are initially close to the truth. Secondly, proponent positions which are close to the truth are comparatively unlikely to be rendered incoherent. This is not simply a finding which derives from the positive relationship between stability and expected verisimilitude reported above; it coheres, moreover, with a theoretical reasoning. All arguments, including those which render proponent positions incoherent, are deductively valid and thence possess at least one false premiss or a true conclusion. That's why an argument can falsify a proponent position only, if it relates to at least one of the position's false truth-value assignments: A false individual conviction of a proponent is taken as premiss of the argument or is itself contradicted by the conclusion. With the arguments being introduced randomly into the debate, proponent positions are therefore more likely to be rendered incoherent by a new argument, the more incorrect truth-value assignments they make, that is, the lower their verisimilitude. The agreement with other proponent positions, in contrast, has no bearing on the likelihood that a proponent position is rendered incoherent by a new argument. These two mechanisms may explain why coincidentally high verisimilitude isn't destroyed by ongoing argumentation, while coincidental consensus is.

Our simulation results indicate that the shape of the SCP—that is, whether the debate is fragmented or compact—affects the verisimilitude dynamics in different ways. A first and somewhat puzzling observation is that compact debates display a higher mean verisimilitude not only throughout the debate but already initially at the very beginning (see Fig. 11.3a). This fact has to be explained by how the debates are categorized into compact and fragmented ones. Recall that we use the aggregated NCC as a measure of compactness. Degree of compactness is, accordingly, determined by (1) the overall shape of the space of coherent positions (and its evolution), as well as (2) how remote, or how central a position the different proponents hold in that space. Now, consider two extreme cases: (a) all proponents are pretty close to the truth and (b) all proponents display a very low verisimilitude. In case (a), the proponents will typically hold rather central positions in the SCP because new arguments are more likely to eliminate positions which are far removed from the truth than those in the truth's vicinity (as detailed above), leaving, as a consequence, a compact center of positions, encompassing the truth, intact. In case (b), however, at least some of the proponent positions will tend to occupy remote positions in the SCP, for new arguments are likely to chop off parts of the SCP which are distant from the truth. So that is why we observe the initial bias: Debates with very high initial mean verisimilitude belong to the very compact debates; debates with very low initial verisimilitude belong to the very fragmented debates. Because of this bias, it is advisable to control for initial verisimilitude when studying the effect of SCP fragmentation. Doing so, we find that the differences in terms of mean verisimilitude evolution become ever more pronounced as the initial verisimilitude decreases (cf. Fig. 11.3b). As regards proponents with high initial verisimilitude, compact and fragmented debates display virtually identical verisimilitude evolutions. When taking off at medium or even low initial mean verisimilitude, however, proponents seem to approach the truth much more steadily and quickly in compact than in fragmented debates. This calls for an explanation. I suggest that the analogies, developed in Chap. 4 in order to understand how the shape of the SCP influences consensus dynamics, might be helpful here, again. We had likened the evolution of a compact SCP to the pulling of a fishing net, where proponent positions correspond to the fishes which are gradually pushed towards each other. With a view to truth dynamics, we have to extend the metaphor. Truth is a position which cannot be displaced and remains fixed. So, assume the fishing net were (not necessarily concentrically) spread around a fishing boat with fixed position and were pulled from that very boat such that the boat always remains within the volume enclosed by the net. The boat, in this analogy, corresponds to the true position. The evolution of fragmented debates, however, has been described by the metaphor of the flooded village. The truth, in this analogy, corresponds to the village's single highest location which represents the sole position that is never flooded—say the castle's tower. With these two extended analogies in mind, we may understand why proponent positions tend to approach the truth much more steadily and quickly in compact debates. For as the village is flooded, its inhabitants might initially resort to elevated locations or buildings other than the castle, increasing their distance to the castle's tower. Only once these unflooded islands—the remaining clusters of coherent positions—are completely flooded will the inhabitants relocate, climbing, eventually, onto the castle's roof and its tower. The fishes caught in the fishing net, however, will be pushed gradually closer and closer to the boat. Figure 11.12 provides an illustration for these two different types of dynamics. In the compact debate (upper half), proponent positions gradually approach the truth, remaining in one and the same component of the SCP as the truth. In the fragmented debate (lower half), however, proponent positions are eventually scattered on different segments of the SCP, some of them remote from the truth. But why are the differences between the two kinds of dynamics more pronounced for proponent positions which are initially very distant from the truth, and why are they barely discernible for proponent positions with high initial verisimilitude? To see this, recall that positions which are close to the truth are less likely to be rendered incoherent by new arguments. Proponents whose initial position exhibits high verisimilitude will, as a consequence, be affected by the argumentation to a lesser degree. In terms of our analogies: Fishes which swim close to the boat that pulls the net, and roofers which are repairing the castle's roof, won't be affected by the shrinking net, respectively, the increasing flood, for a quite long time. And once they are affected by the SCP's shrinking, they tend

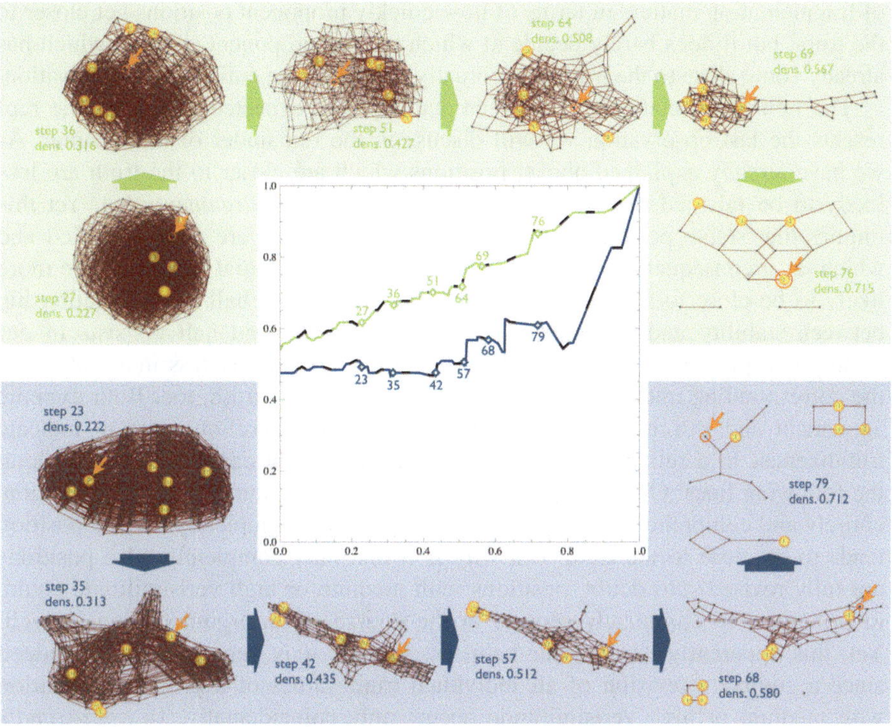

Fig. 11.12 Debate-specific mean verisimilitudes and space of coherent positions for two debates in the ensemble. Mean verisimilitude is plotted against inferential density. The different graphs are 12-dimensional sections of the (20-dimensional) space of coherent positions. The *green curve* and the *upper snapshots* visualize the evolution of a compact debate (top 10th aggregated NCC quantile); the *blue curve* and the *lower snapshots*, in contrast, depict a dispersed debate evolution (bottom 10th aggregated NCC quantile). The time steps of the snapshots of the space of coherent positions are shown in the diagram. The *orange arrows* mark the location of the true position in the space of coherent positions, while the positions occupied by at least one proponent are spotlighted by *yellow circles*

to relocate closer to the truth. Consider, however, a position which is distant from the truth, that is, a fish far apart from the boat or a farmer distant from the castle. While the fish will gradually be pushed towards the boat by the contracting net, the flooding might force the farmer to resort to all sorts of elevated positions (trees, barns, etc.) before he finally ends up on the castle's tower. That is the reason why the different dynamics become apparent only with respect to proponents who are, initially, not very close to the truth. On the basis of these explanations, it is not difficult to understand why compact and fragmented debates give rise to very similar evolutions of the number of completely true proponent positions (cf. Fig. 11.5). The primary difference between compact and fragmented debates relates to parts of the SCP which are distant from the truth, whereas, in the vicinity of the truth, compact and fragmented debates tend to display a similar behavior. So, the degree of fragmentation matters in terms of how quickly proponent positions get closer to the truth, but it does barely decide at which point a proponent position which has already come close to the truth is eventually revised to the entirely correct position.

The nonlinear relationship between stability and expected verisimilitude represents the last observation we will discuss in the remainder of this chapter. As we have already explained above, positions which are closer to the truth are less likely to be falsified in a critical debate with *t-random argumentation*. Yet this implies that stable positions—proponent positions which are rarely falsified and which, as a consequence, stay close to the proponent's initial position—are more likely to be close to the truth. This accounts for the first half of the relationship between stability and expected verisimilitude. The second half consists in our finding that proponent positions which agree by significantly less than 50% with the corresponding initial position tend to be close to the truth, too. Both extreme agreement and extreme disagreement with the initial position seem to indicate truthlikeness in a reliable way. I suggest that this result can be explained along the following lines: Only completely false positions are, during an argumentation, entirely and comprehensively revised. Consequently, a completely revised position tends to be close to the truth. But why is it that only completely false positions are fully revised? No doubt, positions with medium or high verisimilitude could, in principle, be completely revised in the course of an argumentation as well. Yet, this apparently hardly ever happens. In fact, it is very improbable indeed since a complete revision of all individual truth values of a proponent position with medium or high verisimilitude occurs only coincidentally. To *approximate* the corresponding likelihood, consider the simplified situation where a position, defined on 20 sentences, is falsified and readjusted 20 times by altering precisely one randomly chosen truth value at each step. The position hence follows a random walk. At the end of these consecutive modifications, each sentence's truth value (p_i) has been modified a couple of times ($x_i \geq 0$), where the x_i add up to 20. For example, the truth values of sentences $p_1 \dots p_5$ might have been modified four times each, with the other sentences remaining unchanged. There are, altogether, 20^{20}, or ca. $104{,}857 \times 10^{21}$ combinatoric possibilities of how often each sentence's truth value is altered! But in only one single case does the random walk lead to a final position which assigns each sentence exactly the complementary truth value than the original position (viz., the case where each sentence is modified exactly once). This provides at least an idea why it is very improbable that some proponent position will coincidentally—and not because its completely false tenets are gradually corrected by a truth-conducive argumentation—evolve into its complement.

The observed relationship between stability and expected verisimilitude is already discernible at very low densities at which the ensemble-wide mean verisimilitude of proponent positions has hardly changed at all. That is a very important result because it allows one to learn from a controversial argumentation without having to reach high densities. That the relationship vanishes at high densities is therefore of less importance. Besides, this latter observation is not very surprising, either: As *all* positions approach the truth at high densities and as the differences between the proponents, in terms of verisimilitude, gradually disappear, the stability of a proponent position becomes necessarily less significant an indicator of truth.

Chapter 12
The Veritistic Dynamics of Random Debates with Explicit Background Knowledge

Relying on background knowledge in the course of an argumentation allows proponents, or the interpreter who reconstructs a debate, to make use of implicit premisses, which reduces the number of explicit premisses per argument and, consequently, increases the inferential density of the dialectical structure (cf. Sect. 2.5). Accordingly, the rôle of *tacit* background knowledge has already been studied in the previous chapter, when we investigated the truth dynamics at high inferential densities (which can only be reached with implicit background knowledge). In this chapter, we establish background knowledge explicitly by fixing the truth values of a proportion of the debate's sentences in agreement with the correct position $\mathscr{T}$. We will study how the veritistic dynamics depend on the extent of the background knowledge thus introduced, simulating debates with different levels of fixed background beliefs.

12.1 Setup

In analogy to Chap. 5, we consider three levels of basic background knowledge, namely, the background knowledge ratios $\beta = 0.1, 0.2, 0.4$. For each level of background knowledge, we set up an ensemble of 1,000 debate simulations with the corresponding background knowledge ratio and in line with the following specifications:

Argumentation Mechanism: *T-random argumentation* (cf. Sect. 11.1).

Discovery Mechanism: The background knowledge $\mathscr{B}_t$ fixes the truth values of a specific proportion β (namely, 10%, 20%, and 40%) of the n sentence pairs in the sentence pool in a correct way (i.e., the true position $\mathscr{T}$ extends $\mathscr{B}_t$). It remains constant throughout the debate simulation.

Update Mechanism: *Closest coherent with background knowledge* (cf. Sect. 5.1).

Each debate hosts six proponents and terminates once all proponents have acquired the true position.

G. Betz, *Debate Dynamics: How Controversy Improves Our Beliefs*,
Synthese Library 357, DOI 10.1007/978-94-007-4599-5_12,

12.2 Results

The level of explicit background knowledge has a substantial effect on the pace at which proponents approach the truth, as Fig. 12.1 demonstrates. Let's consider the overall mean verisimilitude evolutions first (left-hand panels). The broader the background knowledge, the higher the proponents' initial verisimilitude. More precisely, initial mean verisimilitude equals $(1-\beta)/2$ since, on average, half of a proponent's initial beliefs which don't belong to the background knowledge are true. Besides the initial verisimilitude level, background knowledge seems to affect the subsequent mean verisimilitude evolution, as well. Thus, in the three ensembles, mean verisimilitude increases by roughly 15 percentage points in the density interval 0–0.5. Due to the elevated initial value, verisimilitude increases almost linearly with $\beta = 0.4$: Roughly half of the initially incorrect beliefs have been rectified at $D = 0.5$. In debates with a smaller body of background knowledge, the proportion of corrected beliefs is significantly smaller—a fact being reflected in the upward bending curves. The verisimilitude evolutions of proponents with specific initial verisimilitude (right-hand panels) dovetail nicely with the previous results. As the ratio of background knowledge increases, the verisimilitude evolutions gradually turn into a linear rise. Irrespective of the background knowledge level, however, proponents with medium initial verisimilitude (middle curves) follow, approximately, the debates' mean evolution. Proponents with high initial verisimilitude (top curves) stay close to the truth—verisimilitude is not lost in the course of the controversial argumentation—before closing in on the truth at high densities. Proponents who hold, at first, mainly incorrect positions (bottom curves) catch up with other proponents and exhibit a substantial increase of verisimilitude in the first phase of a debate.

Background knowledge not only influences mean verisimilitude, that is, how close proponent positions are to the truth, but the mean number of proponents who have adopted an entirely true position (verisimilitude = 1), too. While background knowledge seems to have no impact on the initial number of entirely correct proponent positions (compare the left-hand plots in Fig. 12.2), the mean number of correct positions, at $D = 0.5$, equals almost 1 for $\beta = 0.4$ as compared to roughly 0.5 ($\beta = 0.2$) and 0.4 ($\beta = 0.1$). In other words, almost every debate with 40% background knowledge contains a proponent who has found the truth at $D = 0.5$. With 20% (10%) background knowledge, however, this holds merely for every second debate (2 out of 5 debates). Likewise, at the density $D = 0.8$, the number of proponents with fully correct positions in debates with 40% background knowledge is substantially higher (2.6) than in debates with 20% (1.9) or 10% background knowledge (1.5). The right-hand plots of Fig. 12.2 display the mean collapse-to-truth densities of proponent positions with specific initial verisimilitude. As expected, proponents with high initial verisimilitude exhibit comparatively low collapse-to-truth densities, as they adopt the fully correct position at an earlier stage than proponents who possess, at first, more incorrect beliefs. Interestingly, however, background knowledge affects the numerical value of these densities

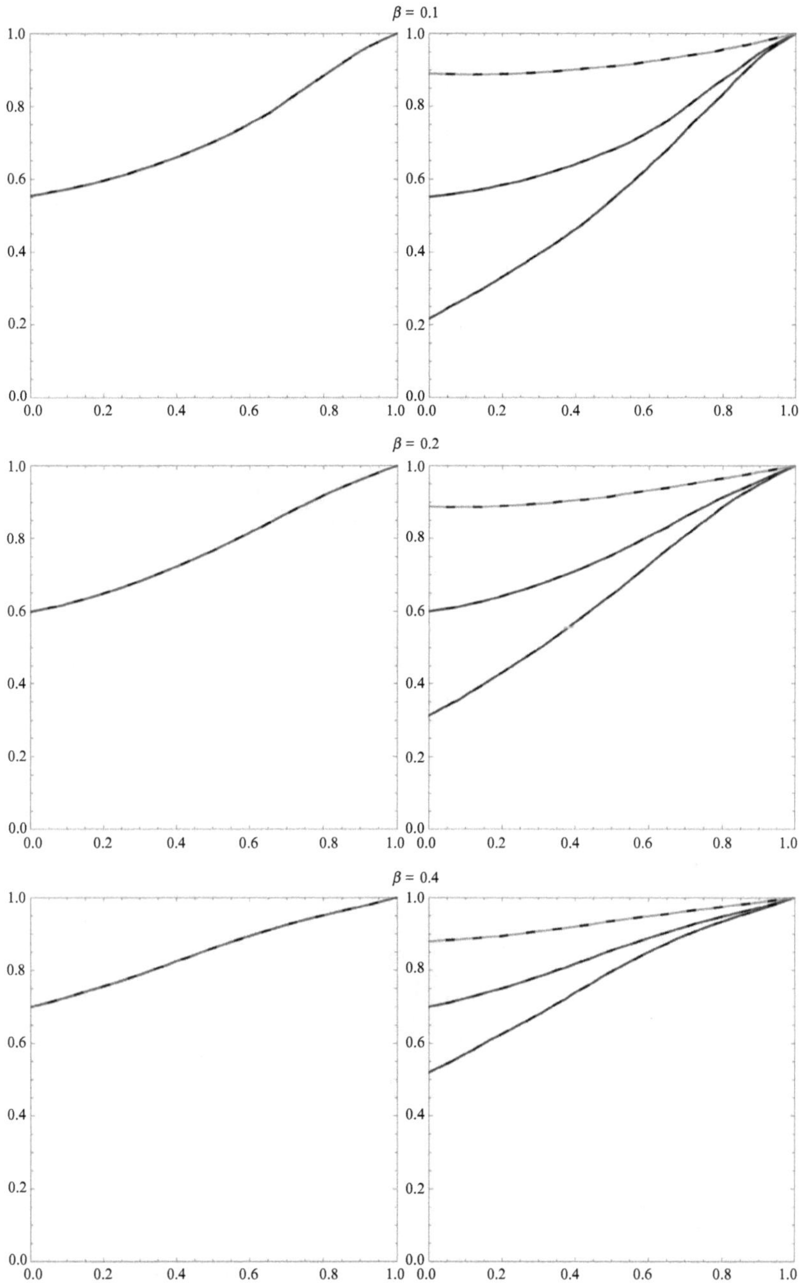

Fig. 12.1 *Left-hand panels*: Ensemble-wide mean verisimilitude evolutions averaged over all proponents and plotted as a function of inferential density. *Right-hand panels*: Ensemble-wide mean verisimilitude evolutions of proponents with different initial verisimilitude; initial verisimilitude intervals based on which the *curves* are calculated depend on β; they are, from *bottom* to *top*, $[\beta;\beta+0.2]$, $[0.4+\beta/2;0.6+\beta/2]$, $[0.8;1]$

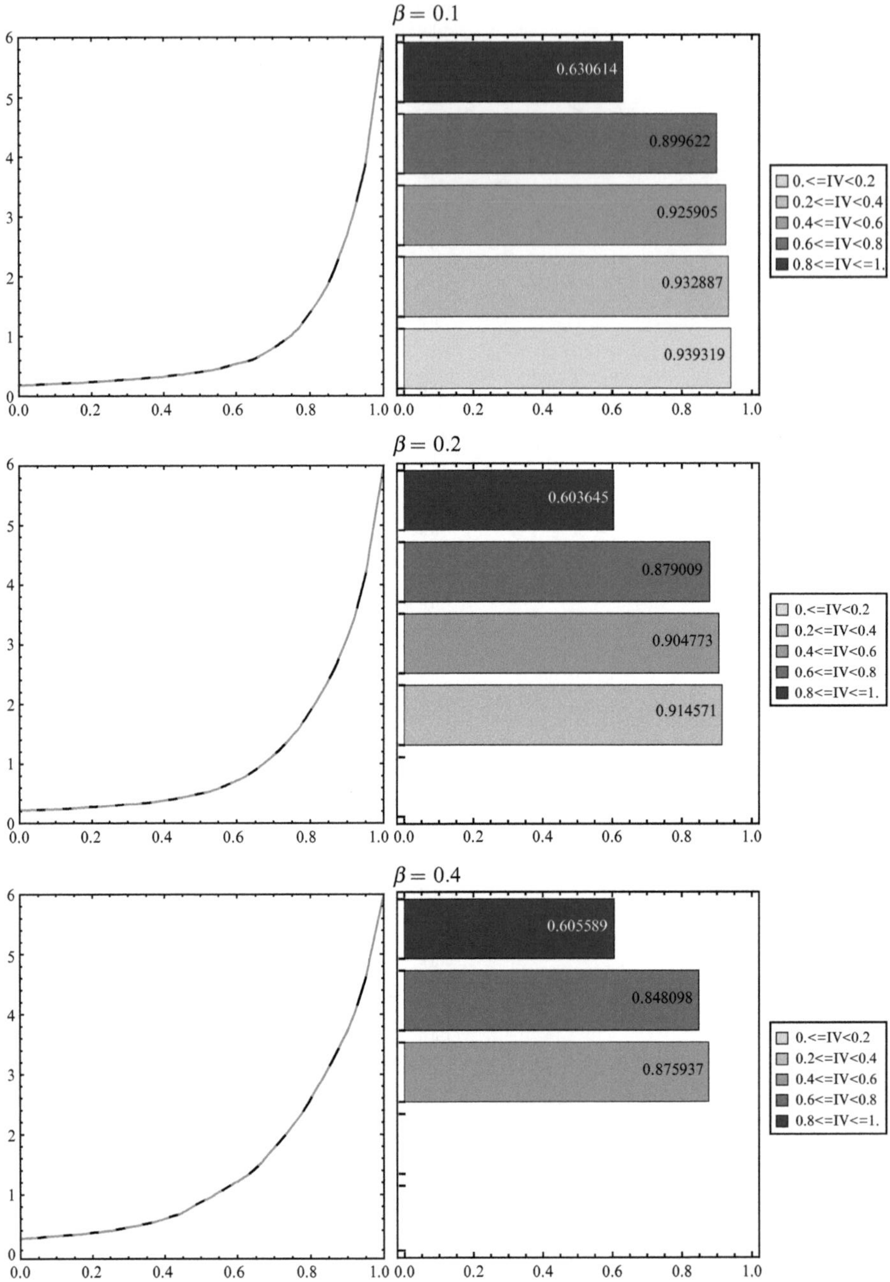

Fig. 12.2 *Left-hand panels*: Ensemble-wide mean number of entirely true positions as a function of inferential density. *Right-hand panels*: Ensemble-wide mean collapse-to-truth densities, that is, mean densities at which proponents with a corresponding initial verisimilitude acquire, on average, the true position

only marginally. For example, proponents with more than 80% true initial beliefs typically acquire a fully correct position at $D = 0.63$ ($\beta = 0.1$), $D = 0.60$ ($\beta = 0.2$), and $D = 0.61$ ($\beta = 0.4$). Background knowledge doesn't help proponents with specific initial verisimilitude to get the full truth any faster. This implies that the number of entirely correct proponent positions increases with broader background knowledge (left-hand panels) merely because background knowledge alters the frequencies at which proponents possess a certain initial verisimilitude. By ruling out that proponents hold completely false initial positions, background knowledge increases the proportion of proponents who adopt initial positions close to the truth. So, with background knowledge, proponents who are already close to the truth don't find the truth any faster than before, yet, they are more numerous than without background knowledge, causing, in sum, the number of fully correct proponent positions to rise more rapidly.

Figures 12.3 and 12.4 display how the veritistic dynamics of highly compact (high aggregated NCC) and highly fragmented (low aggregated NCC) debates deviate from the ensemble mean. As the top left-hand plot in Fig. 12.4 shows, mean verisimilitude takes off at a somewhat higher level and increases slightly more steadily in compact debates (dotted curves) than in fragmented ones (dashed curves) for $\beta = 0.1$. At a low background knowledge ratio, the veritistic dynamics of compact and fragmented debates therefore correspond to the dynamics without background knowledge (see the previous chapter). With higher β, however, this difference starts to blur ($\beta = 0.2$) before it is eventually reversed ($\beta = 0.4$). So, with 20% background knowledge, the mean verisimilitude evolutions in compact and fragmented debates differ only marginally. Yet, with 40% background knowledge, the difference becomes distinct, again: Now, at least for densities between 0.2 and 0.8, proponent positions in fragmented debates are clearly closer to the truth than those in compact debates. This astonishing reversal repeats itself as regards proponents with specific initial verisimilitudes (right-hand plots). At $\beta = 0.1$, proponents with high, medium, and low initial verisimilitude tend to approach the truth more rapidly and steadily in compact than in fragmented debates. This difference vanishes at $\beta = 0.2$ and is reversed at $\beta = 0.4$: Here, proponents in fragmented debates, no matter which initial verisimilitude they exhibit, advance towards the truth at higher pace. Thus, as for the consensus dynamics (cf. Chap. 4), background knowledge reverses the impact of the SCP's fragmentation on the debate's veritistic dynamics.

Background knowledge has, as Fig. 12.4 illustrates, a pronounced impact on how the number of entirely correct proponent positions evolves in fragmented and compact debates, as well. With a small body of background knowledge ($\beta = 0.1$), fragmented and compact debates give rise to very similar evolutions—as is the case for random debates without background knowledge (see Fig. 11.5). With 20% background knowledge, however, fragmented debates host substantially more proponents who have found the truth than compact debates, which agree neatly with the ensemble-wide mean. This clear difference between compact and fragmented debates still holds at $\beta = 0.4$. Yet, as the ensemble-wide mean has risen substantially in comparison to $\beta = 0.2$ (solid curve), compact debates now typically contain a below-average number of fully correct proponent positions.

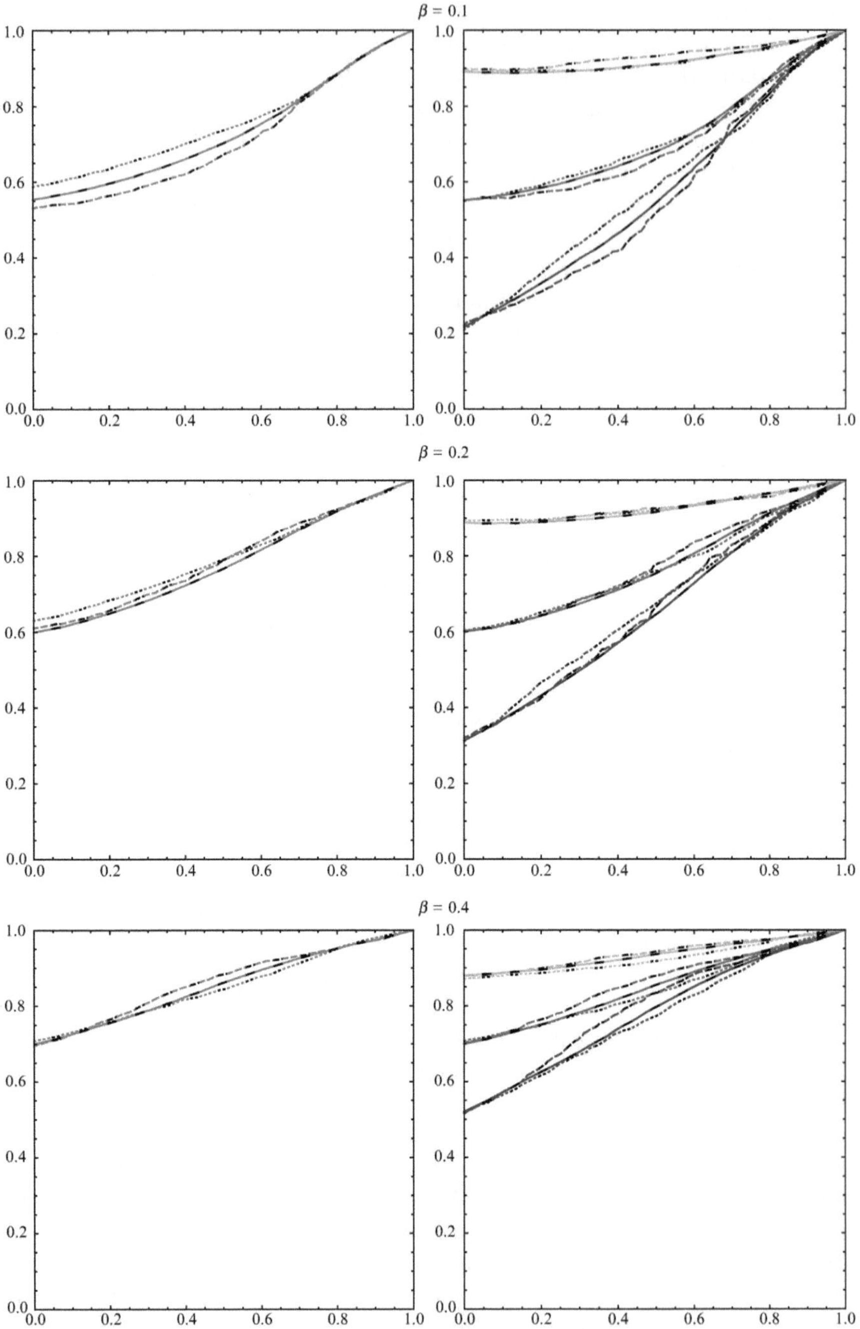

Fig. 12.3 Mean verisimilitude evolutions as in Fig. 12.1. In addition, *dotted curves* (*dashed curves*) display the corresponding verisimilitude evolution in extremely compact (fragmented) debates

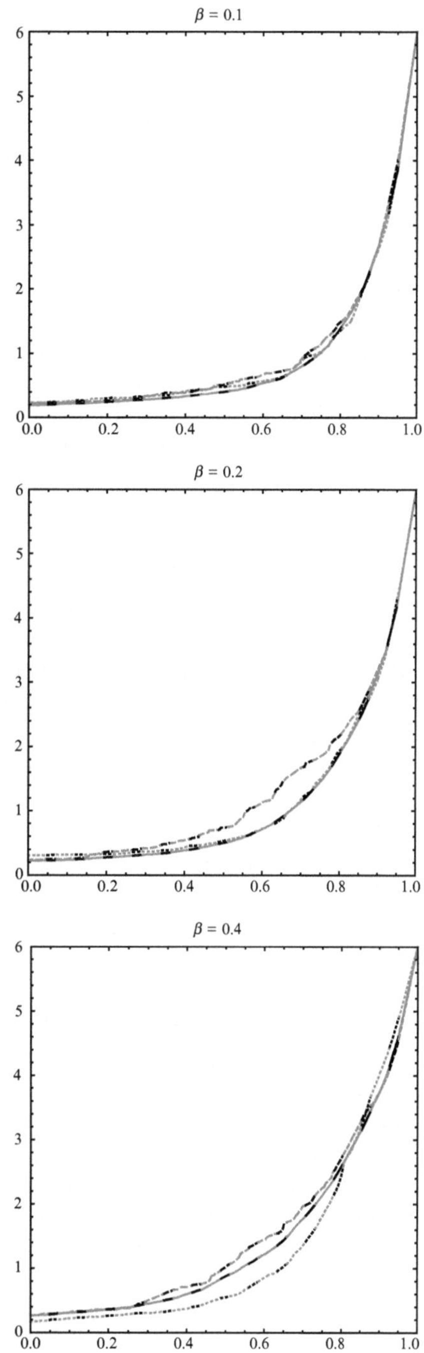

Fig. 12.4 Number of entirely true positions as a function of inferential density. In addition to the ensemble-wide mean—*solid curves*, see Fig. 12.2—*dotted* (*dashed*) curves display the mean evolution in highly compact (fragmented) debates

12.3 Discussion

We will submit two aspects of the previous findings to a more detailed analysis. The first one pertains to the impact of background knowledge on the debates' mean verisimilitude evolutions. As we have seen above, background knowledge increases the truth-conduciveness of argumentation. More specifically, as we broaden the body of background knowledge, a higher proportion of the initially incorrect beliefs gets rectified in a certain density interval, for example, from 0 to 0.5. In the same time, we have found that verisimilitude is, in *absolute* terms, increased by the very same amount no matter how comprehensive the background knowledge—namely, by roughly 15 verisimilitude points at the density $D = 0.5$. Clearly, these observations are consistent because broader background knowledge corresponds with higher initial verisimilitude and, consequently, with a smaller fraction of initially incorrect beliefs. Still, this impact of background knowledge deviates from its effect on the consensus dynamics, where it not merely increased the relative but even the absolute effectiveness of argumentation. It calls hence for an explanation.

The second aspect we will pay closer attention to is the way background knowledge alters the role of fragmentation. Without background knowledge, fragmented debates display a slower and more abrupt verisimilitude increase than compact ones. This still holds if we introduce a limited body of background knowledge ($\beta = 0.1$), yet the impact is reversed at higher levels of background knowledge: Here, proponents in extremely fragmented debates approach the truth more rapidly, and fragmented debates do possess a higher number of entirely true proponent positions as well. We had noticed a similar, albeit much more pronounced, effect with respect to the consensus dynamics. Thus, we will discuss in how far the effect in the veritistic dynamics can be explained along the same lines.

Let us consider the influence of background knowledge on the mean verisimilitude evolutions first! In order to explain the impact of background knowledge on the consensus dynamics, we have introduced, in Chap. 5, the notion of effective background knowledge, $\mathscr{B}_{\text{eff}}$. Besides the basic background knowledge, which consists in the truth values explicitly assigned and fixed in the simulation, the effective background knowledge comprises the truth value assignments which are implied by the basic background knowledge plus the inferential relations encoded in the dialectical structure. If, for example, p_1 and p_2 are true according to the basic background knowledge, which doesn't assign p_3 any truth value whatsoever, and if the debate contains the argument $(p_1, p_2; p_3)$, then p_3 belongs to the effective, yet not to the basic background knowledge. We have derived the following relationship which approximates the proportion of effective background knowledge, β_{eff}, as a function of inferential density (cf. (5.1)),

$$\beta_{\text{eff}} = \left(1 + \frac{D(\tau)}{1 - D(\tau)}\right)\beta.$$

Figure 12.5 plots the mean verisimilitude evolutions of the different ensembles against the evolution of effective background knowledge β_{eff}—as approximated

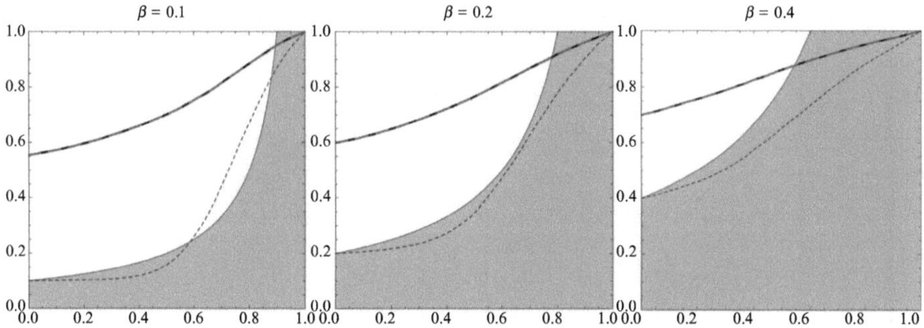

Fig. 12.5 Ensemble-wide mean verisimilitude evolution and ratio of effective background knowledge (β_{eff}) as functions of inferential density. The verisimilitude evolutions correspond to the *left-hand plots* of Fig. 12.1. The ratio of effective background knowledge is plotted according to (a) the analytic approximation (*shaded area*) and (b) the simulation results, which yield an ensemble-wide mean value (*dashed curve*)

by (5.1) (shaded area) and as calculated given the ensemble data (dashed curve). Obviously, (5.1) overestimates the actual effective background knowledge. This contrasts with the ensemble studied in Chap. 5, where the approximation slightly underestimated β_{eff}, and is in need of explanation: We are studying debates with random argumentation and the same level of basic background knowledge. So why don't the levels of effective background knowledge agree? In fact, there exists a small difference concerning the debates' setup which is not as innocent as it appears at first glance. In this chapter's debates, arguments are introduced in line with *t-random argumentation*, that is, a randomly constructed argument is checked for deductive validity before being added to the dialectical structure. Studying the consensus dynamics, however, we didn't impose a true position in the first place, let alone check validity (arguments were simply stipulated to be valid). And this makes a difference. With *t-random argumentation*, a randomly chosen argument is less likely to increase the effective background knowledge than with simple *random argumentation*. More precisely, without including a true position in the simulations, there are twice as many potential arguments which, if introduced into the debate, fix an additional truth value given the basic background knowledge. To see this, consider two sentences, p_1, p_2, which belong to the basic background knowledge. Let p_3 be a sentence whose truth value is not fixed by $\mathscr{B}$. Without taking into account the true position, as the simulations of consensus dynamics do, both $(p_1, p_2; p_3)$ and $(p_1, p_2; \neg p_3)$ represent arguments which might be introduced in line with the *random argumentation* mechanism. However, if we stipulate that there exists a single true, complete position, then both p_1 and p_2, belonging to the background knowledge, are true and, consequently, one of the two arguments has to be invalid. Hence, with *t-random argumentation*, only one of those arguments belongs to the set of potential arguments which may be introduced into the debate. As the validity criterion only eliminates arguments with true premisses from the set of potential arguments, the relative frequency

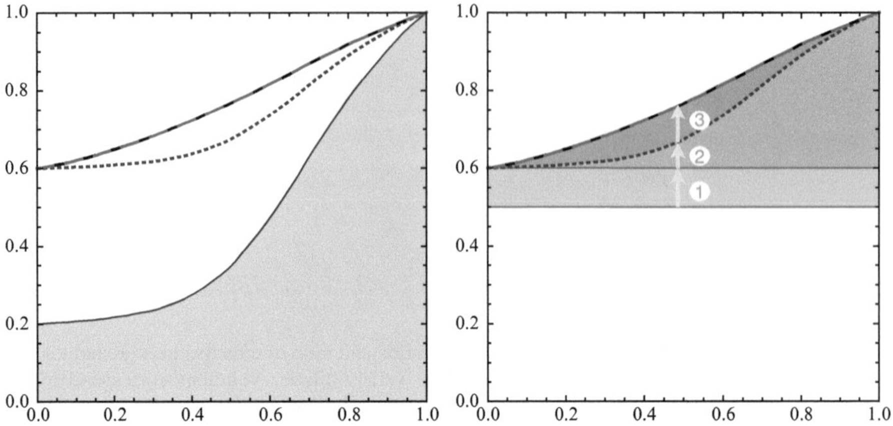

Fig. 12.6 *Left-hand panel*: Expected verisimilitude of randomly assigned (not necessarily dialectically coherent) positions which merely coincide with respect to the effective background knowledge (*dashed*). The precise evolution (not the approximation) of effective background knowledge for $\beta = 0.2$ is plotted as *gray area*, the ensemble-wide mean verisimilitude as *thick curve*. *Right-hand panel*: Illustration of the three mechanisms which lead to truth rapprochement in debates with background knowledge. (*1*) Verisimilitude increases because of basic background knowledge. (*2*) Verisimilitude increases because of effective background knowledge. (*3*) Verisimilitude increases due to the contraction of the space of coherent positions

of arguments that increase the effective background knowledge is significantly reduced with *t-random argumentation*. As a result, the ratio of effective background knowledge increases more slowly in this chapter's ensembles and lies well below the approximation derived in Chap. 5. To understand the veritistic dynamics of debates with background knowledge, we will therefore resort to the observed evolutions of β_{eff}.

In close analogy to our reasoning in Chap. 5, Fig. 12.6 illustrates the three mechanisms by which background knowledge affects verisimilitude. First of all, the basic background knowledge fixes directly some of the proponents' beliefs in a correct way. Proponents may freely assign truth values to only $(1-\beta)$ of the debate's sentences—half of which are, on average, set correctly, given a purely random assignment. This results in an increase of mean verisimilitude by $\beta/2$. Secondly, the very same idea applies not merely to the basic but also to the effective background knowledge. As the inferential density of a debate increases, more and more truth values are fixed, and proponents may assign alternative truth values to less and less sentences. As a consequence, the mean verisimilitude is further increased by $(\beta_{\text{eff}} - \beta)/2$. Finally, a third mechanism consists in the truth rapprochement due to the specific contraction of the space of coherent position. This third mechanism has to account for the residual verisimilitude increase which is not explained by the first two mechanisms. It is also the sole mechanism which drives verisimilitude increase without background knowledge.

Let us put these thoughts together. The effective background knowledge, which gives rise to the second mechanism identified in Fig. 12.6, grows more slowly in the ensembles which examine the veritistic dynamics as compared to those that simulate the consensus dynamics. Accordingly, the second mechanism is less powerful in this chapter's ensembles, relative to those studied in Chap. 5. One and the same level of basic background knowledge thus fosters the effectiveness of argumentation to different degrees. This explains the first aspect discerned at the beginning of this section.

We shall consider the veritistic dynamics of fragmented and compact debates with background knowledge, that is, the second aspect highlighted above, next. Qualitatively, the influence of background knowledge on compact and fragmented debates is the same no matter whether we regard agreement or verisimilitude evolutions: Without background knowledge, agreement and verisimilitude increase more rapidly in compact debates; with (sufficiently broad) background knowledge, however, agreement and verisimilitude increase more rapidly in fragmented debates. Yet, the strength of this reversal is much more pronounced with respect to agreement and consensus evolution. Nonetheless, I suggest that these qualitatively similar observations can also be explained along similar lines. Moreover, the by now well-known analogies of the fishing net and the flooded village may help to explain why fragmentation of the SCP is favorable to verisimilitude increase, albeit to a lesser degree than to agreement increase.

We have explained, in Chap. 5, the superior consensus-conduciveness of argumentation in fragmented debates with background knowledge as follows: Common background beliefs tend to force proponents onto the very same cluster of a fragmented debate. As such fragments can contract more quickly than the entire space of coherent positions, proponent positions typically approach each other more quickly. In terms of the flooded-village analogy: If certain parts of the village are shut off, more inhabitants will assemble on the same building, approaching each other rapidly, as the flood level rises. The proponent dynamics in compact debates, however, are not affected by background knowledge in a comparable way. So, if the fishes caught in a fishing net, for whatever reason, don't enter a certain section of the enclosed volume, this won't speed up their mutual rapprochement as the net is pulled. Regarding truth-conduciveness, essentially similar conditions prevail. Again, background knowledge doesn't fundamentally alter the veritistic dynamics of compact debates. If the fishes don't enter some section of the volume enclosed by the net, this does not lead them to approach the fishing boat at higher pace. However, in a fragmented debate, background knowledge tends to force the proponents on the very same clusters—the inhabitants are more likely to assemble on the same roofs. But, crucially, this speeds up the verisimilitude increase only if it is the castle's roof they resort to. Background knowledge might also compel the inhabitants to flee onto the "wrong" buildings—increasing mean agreement and triggering partial consensus, without fostering (or while even decreasing) mean verisimilitude. So, only if the background knowledge is sufficiently broad to push the proponents towards the cluster that contains the true position, or if that very cluster is itself sufficiently large to fill that part of the SCP which is compatible with the background knowledge, will proponents approach the truth more quickly because of the SCP's fragmentation. This explains why argumentation in fragmented debates becomes significantly more truth-conducive only at a background knowledge ratio of $\beta = 0.4$. Likewise, this reasoning makes the observation intelligible that a sufficiently broad background knowledge is required so that the number of entirely correct proponent positions in fragmented debates exceeds the corresponding number in compact debates.

Chapter 13
Comparing the Veritistic Dynamics of Four Proponent-Specific Argumentation Strategies in Dualistic Debates

We have studied the veritistic dynamics of *random argumentation* (with and without explicit background knowledge) in the previous chapters. In this chapter, we will drop the assumption that arguments are introduced randomly into the debate and suppose that proponents put forward arguments in line with a specific argumentation strategy they pursue. In close analogy to our investigation in Chap. 6, we distinguish and study four argumentation rules: *fortify*, *attack*, *convert*, and *undercut*. We simulate debates with two proponents and examine how the truth-conduciveness of controversial argumentation depends on the strategies chosen by the proponents. In the next chapter, we will extend this analysis to multiproponent debates.

13.1 Setup

We consider the four argumentation strategies: *fortify*, *attack*, *convert*, and *undercut* as defined by Table 6.1. For each pair of argumentation rules (e.g., *fortify–fortify*, *fortify–attack*), we set up an ensemble containing 2,000 debates of the following type:

Argumentation Mechanism: The ensemble-specific pair of argumentation strategies, for example, *convert–attack*, defines the argumentation rules followed by the two proponents in the debate. One proponent implements the first strategy, her opponent the second one. In alternating sequence, the proponents put forward new arguments—one per step—in accordance with their corresponding argumentation strategy: From all potential arguments that satisfy the respective rule and are, in addition, deductively valid (i.e., don't render the true position $\mathscr{T}$ incoherent), the argument to be introduced is chosen randomly. If there is no argument that (i) satisfies a given argumentation rule, (ii) is valid, and (iii) has not

G. Betz, *Debate Dynamics: How Controversy Improves Our Beliefs*, Synthese Library 357, DOI 10.1007/978-94-007-4599-5_13,

been introduced yet, the proponent puts forward a randomly constructed, valid argument.[1]

Discovery Mechanism: The background knowledge $\mathcal{B}$ is empty.

Update Mechanism: *Closest coherent* (cf. Sect. 4.1).

A debate simulation terminates if the two proponents have both settled on the truth or if the inferential density has surpassed 0.8.

13.2 Results

This section gradually unfolds, in more and more detail, the complex results of the debate simulations. To start with, we consider Fig. 13.1, which displays the ensemble-wide mean verisimilitude evolutions in the style of Chap. 6. It suggests that the *undercut* rule is the most truth-conducive argumentation strategy. The top four ensembles regarding mean verisimilitude increase (plots at bottom row) contain at least one proponent who implements the *undercut* strategy. This entails, in particular, that choosing *undercut* as an argumentation strategy is optimal with respect to maximizing mean verisimilitude.[2] Moreover, the highest mean verisimilitude is attained when both proponents implement the *undercut* rule, in which case more than 75% of the proponents' individual beliefs are, at a density $D = 0.5$, correct, as compared to slightly less than 65% for a *random argumentation*.

As regards the homogeneous debates, that is, the debates where the two proponents implement identical strategies and which lie on the figure's diagonal, *convert* is only marginally more truth-conducive than *fortify* and substantially less so than *undercut*. This stands in stark contrast to the rules' relative performance in terms of consensus-conduciveness (cf. Chap. 6). The *attack* rule, however, represents the least effective strategy—both with regard to truth- and with regard to consensus-conduciveness.

The picture becomes more complicated if we consider the inhomogeneous debates as well. Some constellations such as *fortify–attack* or *fortify–convert* yield mean verisimilitude evolutions which are somehow similar to the evolutions we observe when both proponents pursue either the first or the second strategy. Yet other combinations of rules give rise to a rather surprising dynamic. Most remarkably, debates with one attacking and one converting proponent exhibit much higher

[1]It is because of this fallback option to *random argumentation* that some verisimilitude evolutions in Figs. 13.1–13.6, such as *fortify–fortify* in Fig. 13.1, display a sudden and sharp upward turn at densities close to 0.8. This sudden rise indicates the point where there exist typically no more arguments that can be introduced in line with the corresponding strategy and where randomly constructed arguments spawn new inferential relations.

[2]That is, the constellation *undercut*–$rule_1$ causes mean verisimilitude to increase at least as much as (and often substantially more than) any other constellation $rule_2$–$rule_1$ does, for arbitrary rules $rule_1$ and $rule_2$.

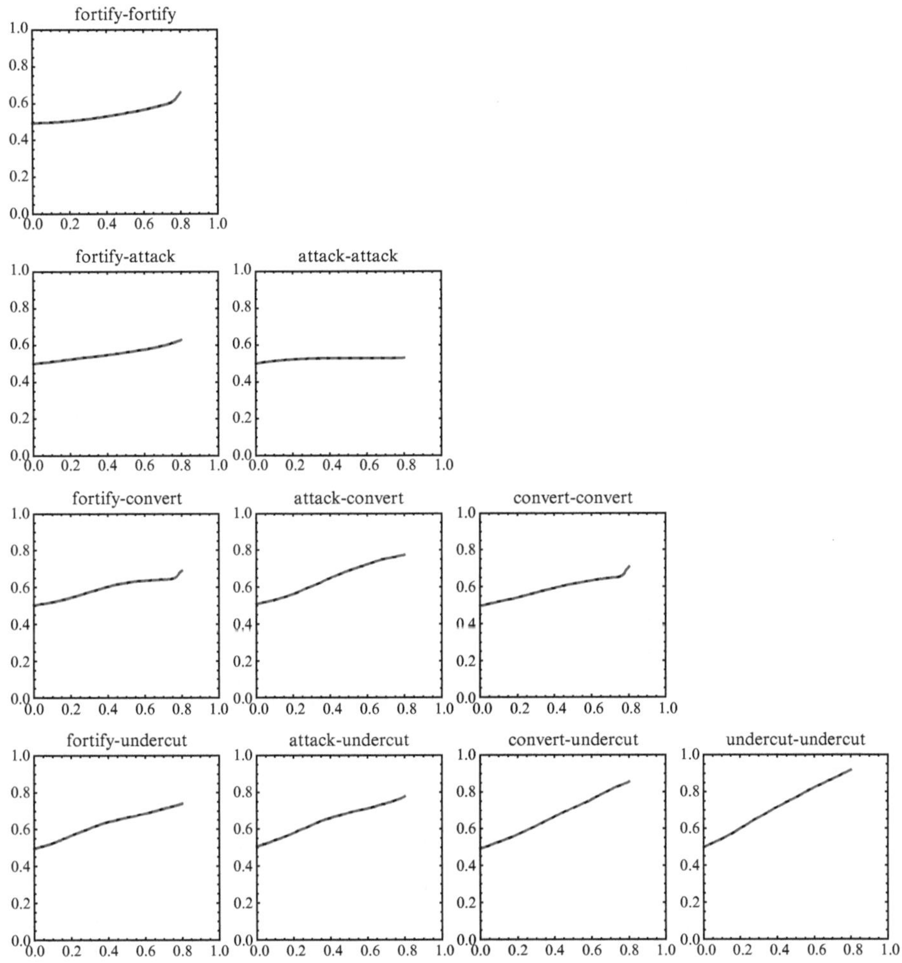

Fig. 13.1 Ensemble-wide mean verisimilitude evolutions, as functions of inferential density, for ten ensembles with two proponents each who pursue the argumentation strategies indicated on top of the diagrams

truth-conduciveness than the *convert–convert* or the *fortify–convert*, let alone the *attack–attack* constellation! The veritistic value of the individual argumentation rules seems to depend substantially on the dialectic context in which they are employed.

Figure 13.2 displays how the number of proponents with a fully correct position increases in our ten ensembles. It confirms the general observations we have made so far. Thus, proponents acquire a fully correct position more rapidly in debates where at least one proponent implements the *undercut* strategy. The *undercut–undercut* combination outperforms all other constellations. The difference

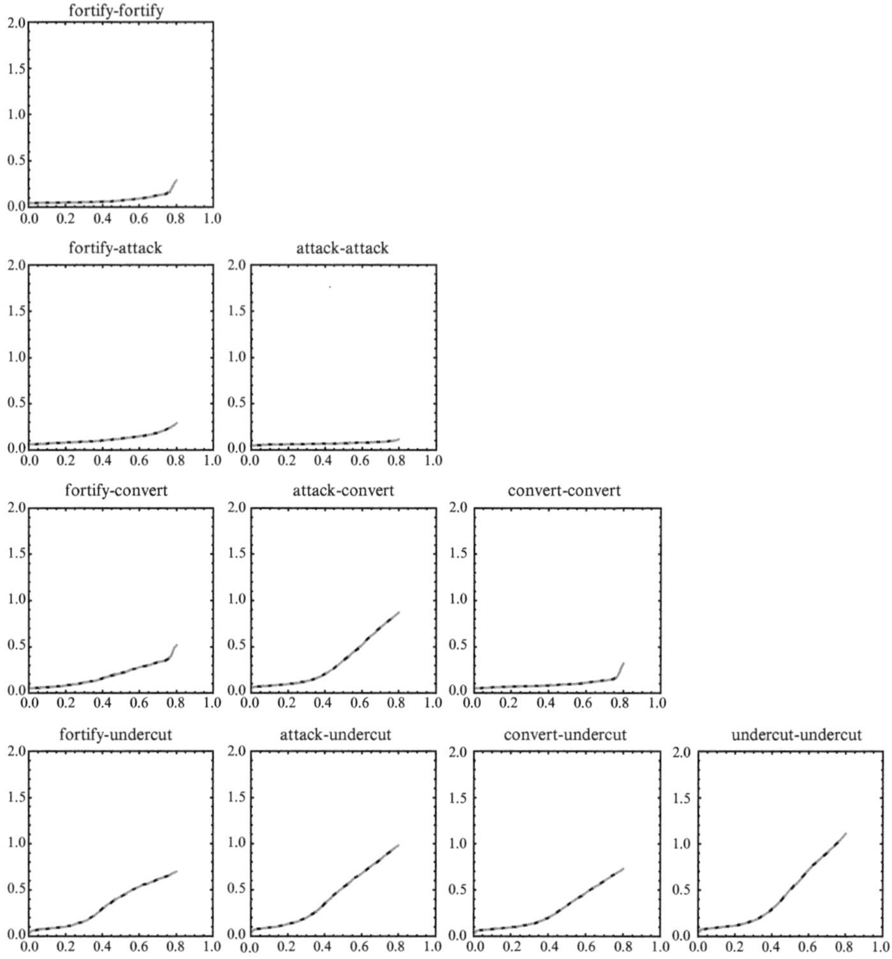

Fig. 13.2 Ensemble-wide average number of entirely true proponent positions, as function of inferential density, in the ten ensembles, plotted as a function of inferential density

between *convert–convert* and *undercut–undercut* is particularly striking. Finally, the constellation *attack–convert* displays the third best truth-conduciveness in terms of the number of proponents with a fully correct position, in spite of the fact that *attack–attack* yields the poorest performance.

With a view to obtaining a more detailed picture of the veritistic dynamics, we have to note that two proponents in one debate may display very different individual verisimilitude evolutions. This is a major difference to the consensus dynamics, where the first proponent agrees with the second one to the same extent as the second with the first one. Because of this symmetry, proponents in dualistic debates cannot exhibit different proponent-specific agreement evolutions. But they may very well exhibit different proponent-specific verisimilitude evolutions. And this pinpoints the first issue we will study in the following: Do proponents with specific argumentation strategies possess systematically different verisimilitude evolutions in one and the same debate? On the basis of this investigation, we will subsequently uncover more details of the veritistic dynamics by distinguishing proponents with different initial verisimilitude and debates in which the two proponents agree, initially, to a very high or low degree.

Figure 13.3 plots the proponent-specific verisimilitude evolutions derived from the ten ensembles and thence addresses the question how rapidly a proponent who (i) implements such and such a strategy and who (ii) faces such and such an opponent approaches the truth. This display of proponent-specific verisimilitude evolutions demonstrates that the effectiveness of a proponent's effort to track the truth primarily depends on the argumentation strategy pursued by her opponent and not the one followed by herself. Thus, proponents who oppose the *fortify* or the *attack* rule (i.e., face, more precisely, an opponent implementing one of these rules) display a comparatively weak increase of verisimilitude (left-hand columns of Fig. 13.3). Proponents, in contrast, whose opponent follows the undercut strategy, approach the truth at considerably higher pace (right-hand column).

In homogeneous debates (diagonal), the proponent-specific verisimilitude evolutions don't, of course, deviate from the debate-wide mean. As the opponent's strategy largely determines the proponent specific verisimilitude evolution, we observe, however, significant differences between the proponents' veritistic dynamics in inhomogeneous debates. Consider in particular the debates with the *fortify* or *attack* rule on the one side and the *convert* or *undercut* rule on the other side (upper-right and bottom-left quarter). Proponents who implement *convert* or *undercut* (and are opposed by *fortify* or *attack*) approach the truth at substantially lower pace than their opponents.

Figure 13.4 displays proponent-specific verisimilitude evolutions while distinguishing proponents with high, medium, and low initial verisimilitude. In all constellations, proponent positions with low initial verisimilitude approach the truth (sometimes significantly) more rapidly than proponent positions with medium or high initial verisimilitude. This dovetails with our findings in previous chapters and is not surprising.

For proponent positions with both medium and low initial verisimilitude, the speed at which their verisimilitude increases during a debate grows as one moves in the table of plots from left to right and is hence substantially influenced by the corresponding opponent's argumentation strategy. Proponent positions with high initial verisimilitude, however, display a contrary behavior: If the opponent implements *fortify* or *attack*, the verisimilitude tends not to change (constellations at the left-hand side of the matrix); yet if the opponent follows the *convert* or, in particular, the *undercut* rule, the initially close proximity to the truth typically evaporates and the proponents recuperate their lost verisimilitude only at high densities. In some constellations—namely, *attack–fortify*, *attack–attack*, and *undercut–attack*—the coincidentally high initial proximity to the truth is not merely provisionally lost but systematically destroyed throughout the debate without ever being regained. In sum, the opponent's strategy largely determines the dynamic of proponent positions with high initial verisimilitude too.

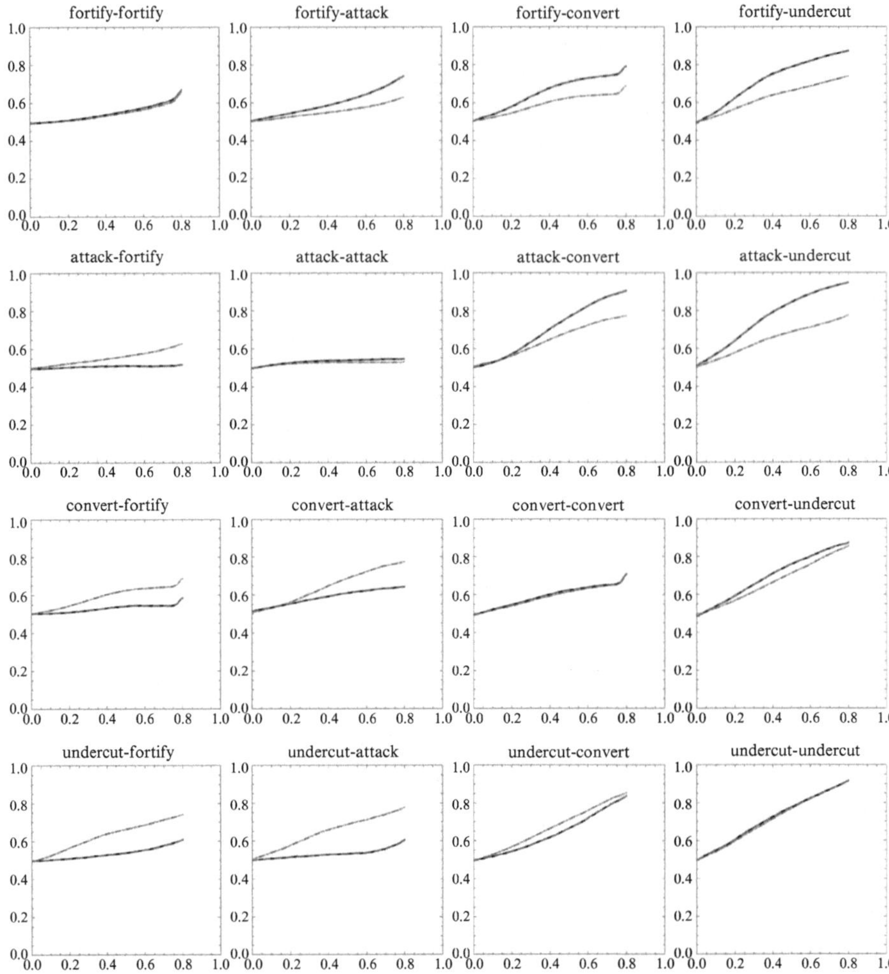

Fig. 13.3 Ensemble-wide mean proponent-specific verisimilitude evolutions, as functions of inferential density, for this chapter's ensembles. A plot labeled "rule_1–rule_2" displays the verisimilitude evolution (*dark curve*) of a proponent who follows $rule_1$ and is opposed by an opponent who adopts $rule_2$. Accordingly, the rows of this table of diagrams correspond to the strategy pursued by the proponent; the columns sort the plots in line with the opponent's strategy. Plots labeled "rule_1–rule_2" and "rule_2–rule_1" are thus derived from one and the same ensemble. Mean verisimilitude evolutions as averaged over the debates' two proponents, shown in Fig. 13.1, are plotted as *light curves* and may serve as a point of reference

So far, our results have taken account of the proponent's and opponent's argumentation strategies and the initial verisimilitude of the proponent. Besides the initial distance to the truth, however, a proponent's distance to her opponent may also bear on the debate's veritistic dynamic. This is at least suggested by the

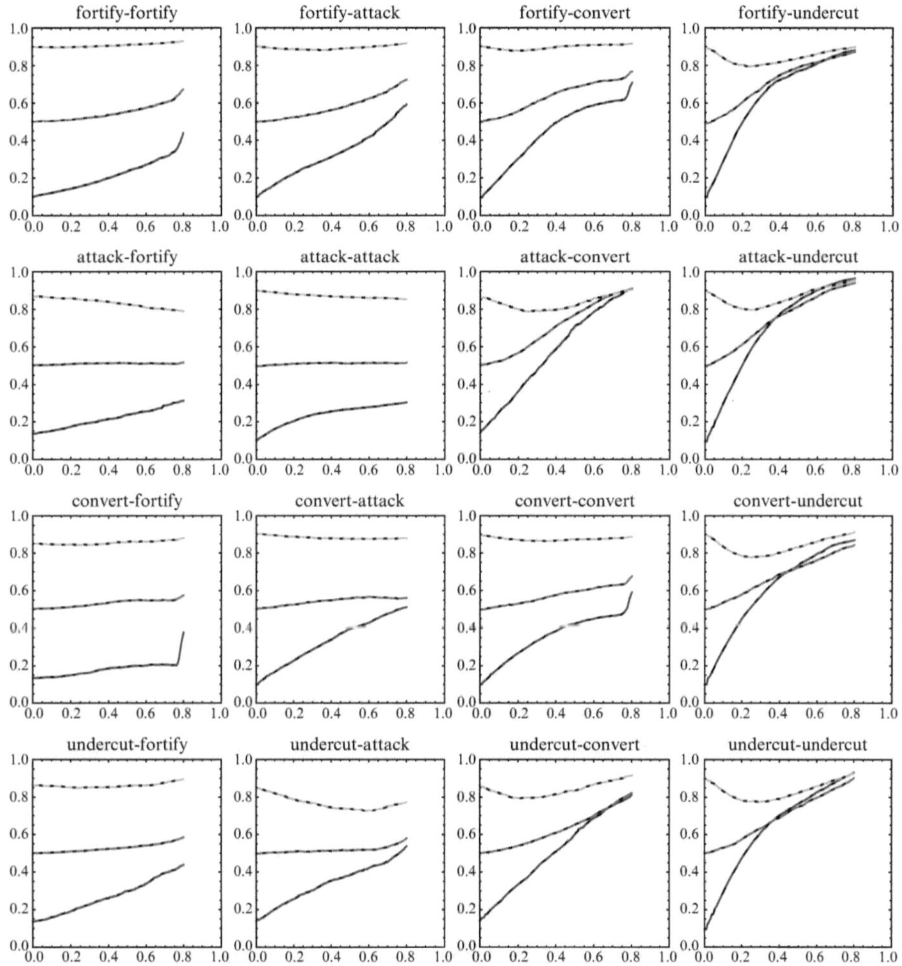

Fig. 13.4 Ensemble-wide mean proponent-specific verisimilitude evolutions, as functions of inferential density, for this chapter's ensembles. In contrast to Fig. 13.3, the plots distinguish proponent positions with different initial verisimilitude, namely, high (0.8–1, *top curves*), medium (0.4–0.6, *middle curves*), and low (0–0.2, *bottom curves*)

fact that initial agreement clearly influences the consensus dynamics of debates with purposive argumentation strategies (cf. Chap. 6). Moreover, initial agreement and initial verisimilitude are, obviously, independent factors: A proponent may, for instance, possess mainly correct beliefs, yet disagree substantially with her opponent, or she may have reached a consensus with her opponent while holding a completely false position. Thus, besides distinguishing proponents with different initial verisimilitude, we will, in the following, classify debates according to the two proponents' initial agreement.

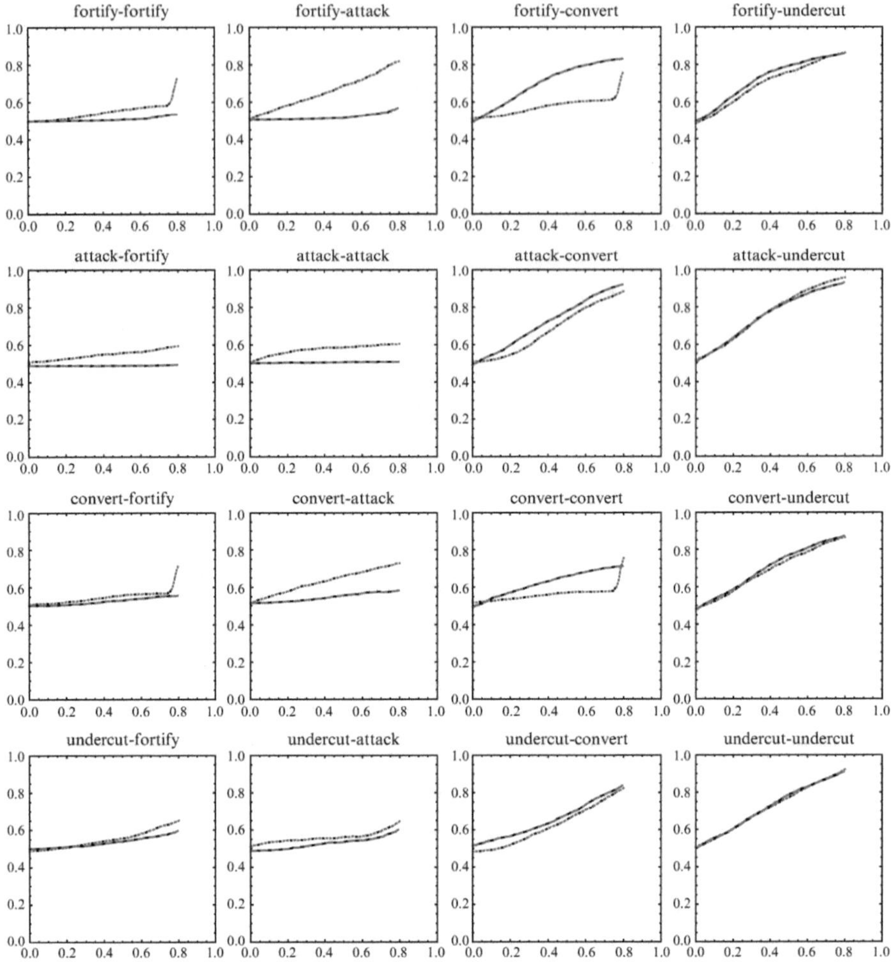

Fig. 13.5 Ensemble-wide mean proponent-specific verisimilitude evolutions, as functions of inferential density, in debates with varying degrees of initial mean agreement. Verisimilitude evolutions are not calculated with respect to all debates in the corresponding ensemble (as in Fig. 13.3). Instead, *dotted curves* show the verisimilitude evolutions in debates where the two proponents agree initially to a large extent ($0.8 \leq (1-\Delta) \leq 1$), and *dashed curves* (which may, due to technical shortcomings, look like *solid ones*) display the verisimilitude evolutions in debates with high initial disagreement ($0 \leq (1-\Delta) \leq 0.2$)

Figure 13.5 depicts the impact of initial agreement on the veritistic dynamics by plotting ensemble-wide mean verisimilitude evolutions (averaged over all proponents, irrespective of their initial verisimilitude) for debates with high (dotted) and low (dashed/solid) initial agreement. Evidently, initial agreement among the proponents has a significant influence on the veritistic dynamics in some ensembles. Still, the specific effect varies from constellation to constellation. To see this, we focus on four settings (ensembles) where the difference is most pronounced, leaving in particular the ensembles with *undercut* aside, where initial agreement has virtually no effect on the veritistic dynamics.

Let us consider, first of all, *fortify–attack* and *fortify–convert*. In both ensembles, the proponent implements the *fortify* rule, while being opposed by the *attack* strategy in the first case and the *convert* strategy in the latter. This makes, apparently, a huge difference. Facing an opponent who follows the *attack* rule, the fortify proponent benefits, in terms of truth-conduciveness, from broad initial agreement. More specifically, if the fortify proponent agrees, initially, by more than 80% with the attack opponent, her verisimilitude increases significantly throughout the debate (dotted curve)—if, on the contrary, initial agreement is minimal (≤ 0.2), the fortify proponent doesn't approach the truth at all. This relationship turns upside down if the opponent implements the *convert* rule: Here, initial *dis*agreement fosters truth-conduciveness. And a fortify proponent who initially agrees with her convert opponent does hardly track the truth at all.

Consider two further ensembles: *convert–attack* and *convert–convert*. They exhibit, by and large, the same pattern as the previously examined constellations. Initial agreement between proponent and opponent fosters the verisimilitude increase of a convert proponent who faces an attack opponent. If both, proponent and opponent, follow the *convert* strategy, however, initial agreement is detrimental in terms of truth-conduciveness.

Extending our perspective to the other ensembles, we may observe that high initial agreement tends to promote the capacity to track the truth of proponents who face the *fortify* and, specifically, the *attack* rule (two left-hand columns of plots). If a proponent, however, is opposed by the *convert* strategy (third column), initial consensus turns out to be detrimental, and the proponent approaches the truth at higher speed in case initial agreement is very low.

Figure 13.6 spells out in more detail when and where high initial agreement between the proponents fosters (or obstructs) their verisimilitude increase by plotting verisimilitude evolutions for different initial agreement *and* initial verisimilitude levels.

Consider the four constellations we have investigated more closely in the context of Fig. 13.5. As the plots clearly demonstrate, the initial agreement among the proponents exerts the greatest influence on the veritistic dynamics if the corresponding proponent possesses a mainly incorrect initial position (bottom curves). If, for example, a fortify proponent with a low initial verisimilitude agrees largely with her attack opponent (bottom dotted curve in *fortify–attack*), she has corrected half of her false initial beliefs at a density of $D = 0.5$ (increasing her verisimilitude from 0.1 to 0.55). Yet, in case she disagrees with her opponent (bottom dashed curve in *fortify–attack*), her verisimilitude is not altered at all. Likewise, fortify proponents whose opponents follow the *convert* rule exhibit a similarly stupendous rise in verisimilitude if they hold a predominantly false initial position and disagree originally with their opponents to a large extent (bottom dashed curve in *fortify–convert*). They correct, however, only a small fraction of their false beliefs if they shared initially most of their beliefs with their convert opponents (bottom dotted curve in *fortify–convert*).

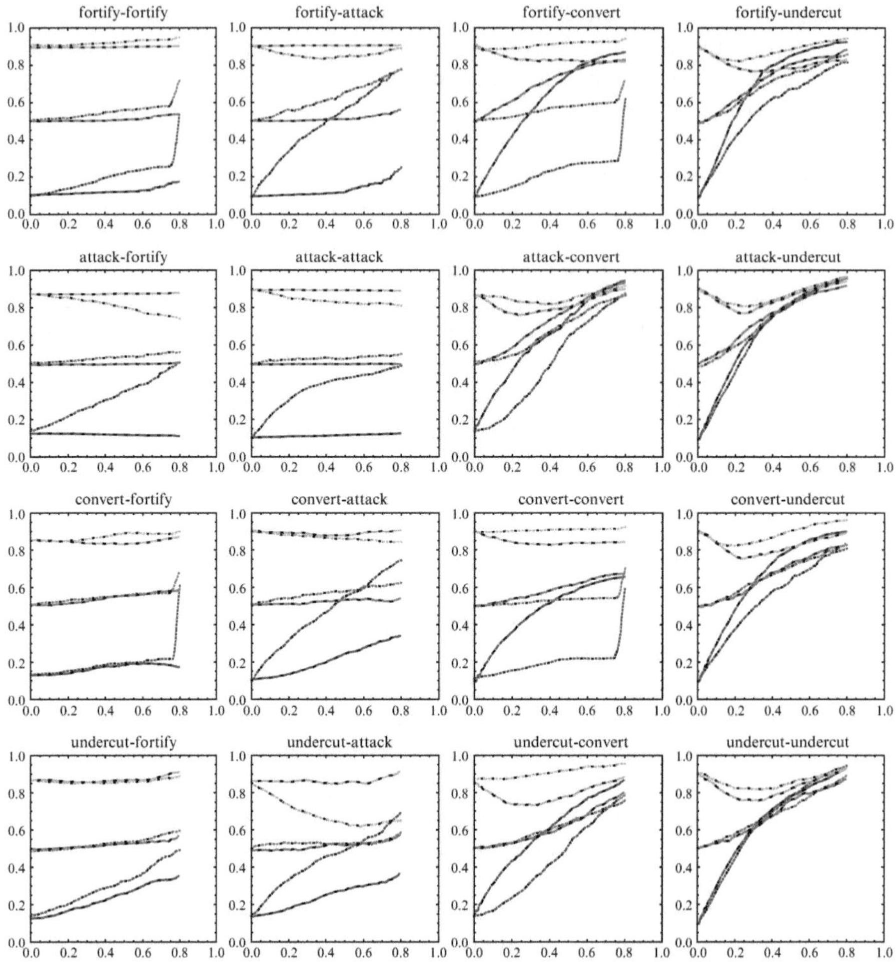

Fig. 13.6 Ensemble-wide mean proponent-specific verisimilitude evolutions, as functions of inferential density, of proponents with different initial verisimilitude and averaged over debates with varying degrees of initial mean agreement. Combining the distinctions introduced in Figs. 13.4 and 13.5, the plots in this figure display the verisimilitude evolutions of proponents with high (*top*), medium (*middle*), and low (*bottom*) initial verisimilitude in debates where the proponent agrees, in the beginning, largely (*dotted*) or marginally (*dashed/solid*) with her opponent

The detailed plots also reaffirm our previous observation (cf. Fig. 13.5) that the initial agreement is least influential with regard to proponents faced by an opponent who pursues the *undercut* strategy.

Finally, the diagrams reveal a general pattern. In most constellations, high initial agreement helps proponents with high initial verisimilitude to get closer to the truth if it prevents proponents with low verisimilitude from doing so. And, vice versa, high initial agreement helps proponents with low initial verisimilitude to track down truth if it obstructs verisimilitude increase for proponents with high initial verisimilitude. (In terms of the diagrams, the top dotted curves lie above the top dashed ones if and only if the bottom dotted curves are located underneath the bottom dashed ones.) This general pattern is realized most markedly in the constellations *attack–fortify*, *fortify–attack*, *attack–attack*, and *undercut–attack* on the one side, where initial consensus diminishes truth-conduciveness given high initial verisimilitude, as well as in the constellations *fortify–convert*, *attack–convert*, *convert–convert*, *undercut–convert*, and *convert–undercut* on the other side, where initial consensus promotes truth-conduciveness given high initial verisimilitude.

13.3 Discussion

The previous section has uncovered the rather intricate veritistic dynamics of dualistic debates with different argumentations strategies. I suggest that the diverse verisimilitude evolutions, which we have observed above, can be explained by reference to the following four distinct, though not entirely independent factors:

1. The consensus-conduciveness of the argumentation strategies employed by the proponent and her opponent
2. The likelihood that the opponent's strategy leads to a falsification of the proponent's position (i.e., the "aggressiveness" of the opponent's strategy)
3. The initial verisimilitude of the proponent
4. The initial agreement between the debate's proponent and opponent

These four factors largely determine a proponent's ability (a) to generate the right kind of agreement, while avoiding (b) spurious consensus.

In this section, we will explain the reported results in reverse direction, starting with the last and most detailed diagrams (Fig. 13.6). Understanding the veritistic dynamics in depth will prove key to explaining the more aggregate findings reported at the beginning of Sect. 13.2.

On the background of the four central factors just identified, we are in a position to discern different types of dialectic constellations. In particular, we may distinguish four general constellations that are characterized, respectively, by high or low initial verisimilitude, as well as by high or low initial agreement. These situations display, as we will see below, specific veritistic dynamics, which largely depend on the consensus-conduciveness of the argumentation strategies employed. That's why our earlier findings about the different strategies' consensus-conduciveness (Chap. 6) turn out to be crucial for understanding the veritistic dynamics. By subsuming the concrete ensembles studied in this chapter under the generally characterized types of constellations, we may thus explain the results presented in the preceding section.

Hence consider, first of all, a proponent with a high initial verisimilitude facing an initially close opponent. Under these conditions, the opponent, too, holds a position close to the truth. That the opponent pursues, in such a situation, a consensus-conducive strategy (such as *convert*) is clearly beneficial in terms of truth-conduciveness since it makes the proponent stick to her coincidentally correct position. If, however, the opponent pursues a strategy which tends to obstruct or even demolish consensus (such as *attack*), proponent and opponent will be driven apart, and the proponent's initially high verisimilitude will be reduced.

Consider, as a second constellation, a proponent initially close to the truth, yet far apart from her opponent. The opponent position displays, consequently, a low verisimilitude. In this case, the employment of a highly consensus-conducive argumentation strategy by the opponent becomes detrimental in terms of truth-conduciveness. For this merely pulls the proponent away from her coincidentally correct position towards the opponent's largely incorrect one; it generates a flawed and spurious consensus. If, in contrast, the opponent argues so as to increase mutual disagreement, the proponent will benefit in the sense of being kept at distance or being pushed away from a largely false position; she will, accordingly, retain her high verisimilitude.

These two constellations just described neatly explain the differences we observe when comparing the various top curves in the *attack* and *convert* columns in Fig. 13.6.

Next, let us picture a proponent holding an initial position which is mainly incorrect and close to her opponent's initial position. Thus, the opponent's position possesses a low verisimilitude as well. Assume, in a first step, the opponent argues in a consensus-conducive way, strengthening the mutual agreement with the proponent rapidly and effectively. This will obviously cause them to stick to their mainly false positions and prevent them from tracking down the truth. In this case, consensus-conducive argumentation leads to and consolidates spurious consensus. But now assume, in contrast, that the opponent employs a strategy which reduces mutual agreement. As a result, the opponent will push away the proponent, who will be forced to give up her more or less completely false position and thus gradually approach the truth. An argumentation strategy which is detrimental in terms of consensus-conduciveness benefits the proponent in terms of truth-conduciveness.

Finally, consider a constellation with low verisimilitude of the proponent's initial position and low initial agreement between proponent and opponent. Hence, the opponent's initial position is, to a large extent, correct. If the opponent argues in a consensus-conducive way, she will effectively pull the proponent towards her (mainly correct) position, thereby increasing the proponent's verisimilitude. Consensus-conducive argumentation is truth-conducive. But assume the opponent argues so as to obstruct a mutual rapprochement with the proponent. This prevents the proponent from approaching the opponent, and hence the truth, and is highly detrimental in terms of truth-conduciveness (as regards the proponent's position).

These last two constellations we have just contemplated provide a sound explanation for the differences regarding the bottom curves in the second and third column of Fig. 13.6.

Let us now move on to more general facts to be explained. It is clear that in debates where the proponent's position is hardly ever rendered dialectically incoherent (typically because the opponent follows the *fortify* or the attack rule and the initial mutual agreement is medium-sized or low), the proponent doesn't substantially approach the truth since she's not compelled to modify her position in whatever direction. This accounts for the relatively flat verisimilitude evolutions on the left-hand side in Figs. 13.3–13.6.

Moreover, the notion of spurious consensus, which prominently figured in some of the special constellations described above, renders further peculiar observations intelligible. Thus, a combination of *convert* and *attack*, as opposed to *convert–convert*, helps to preempt spurious consensus (since argumentation is, given such a combination, much less consensus-conducive) and thence performs better in terms of truth-conduciveness. This holds even, though to a lesser degree, for *undercut–undercut*: In such debates, too, consensus might be spurious and that's why an attack proponent who obstructs consensus and who is faced by an undercut opponent tracks down truth even more effectively than an undercut proponent facing the *undercut* strategy.

I suggest that the relative differences in terms of truth-conduciveness between the argumentation rules can be explained along the lines just sketched. Still, there remains a final question which calls for a different type of explanation. Why does the *undercut* strategy lead to such a rapid increase of verisimilitude in general, specifically in comparison to a *random argumentation*? Or, more precisely, why is frequent falsification beneficial in terms of truth conduciveness at all? Why doesn't rendering a proponent's position dialectically incoherent compel her to move away from the truth or to oscillate in the space of coherent positions without approaching the truth more rapidly than in the case of *random argumentation*?

A first part of an answer to this question consists in recalling that frequent falsification is not beneficial in terms of truth-conduciveness as regards positions with high initial verisimilitude (cf. Fig. 13.4, right-hand column). Proponents who are coincidentally close to the truth tend to lose verisimilitude when facing an undercut opponent and don't recuperate that lost proximity to truth any faster than their counterparts in, for example, random debates. Yet this fact already implies that the superior effectiveness of the *undercut* strategy, or of frequent falsification, to put it more generally, must stem from its performance regarding initial positions with medium and low verisimilitude. Now it isn't surprising at all that completely false positions benefit from being rendered incoherent frequently because such positions cannot be modified but to the better. Yet why do regular falsifications tend to improve positions with medium verisimilitude, for example, with 50% incorrect individual beliefs? Why do proponents with such convictions, when compelled to modify their position, tend to give up false rather than correct beliefs? Given the purely random *closest coherent* update mechanism, we would expect them to rectify a false belief as often as to abandon a true one.

To solve this riddle, it is helpful to reframe it. Rendering a proponent's position incoherent by putting forward an argument (with two premisses) is equivalent to revealing an internal inconsistency, that is, one identifies three sentences each of which the proponent considers, incoherently, true, thence forcing the proponent to give up at least one of them. The only restriction on constructing this kind of internal paradox consists in the fact that at least one of the three sentences is objectively false

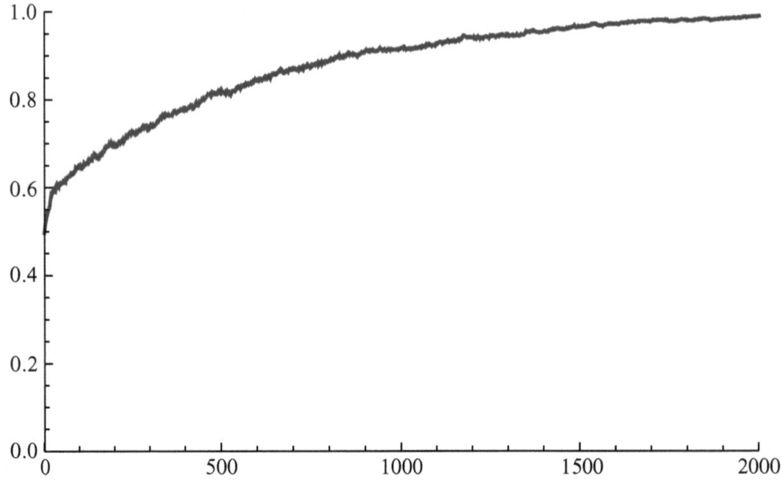

Fig. 13.7 Verisimiltude increase in a highly simplified model of dialectic falsification. At each step, a proponent is required to alter one of three randomly selected beliefs (not all of which are true). The plot displays the mean verisimilitude evolution (y-axis) averaged over 500 proponents for the first 2,000 steps (x-axis)

(otherwise the corresponding argument is not valid). But while the three beliefs, at least one of which the proponent has to abandon, must not all be true, they may very well all be false. And that is the reason why, on average, incorrect beliefs are more likely to figure in the inconsistencies presented to the proponent than correct beliefs. Consequently, proponents whose positions are constantly rendered incoherent are more likely to modify some of their incorrect rather than their correct beliefs.

This admittedly abstract explanation can be nicely illustrated by a very simple model. We consider a proponent who holds 20 individual beliefs. The correct position $\mathcal{T}$ as well as the proponent's initial position is determined randomly. At each time step, we choose three random beliefs of the proponent—not all of which are true—and, henceforth, modify the proponent's position with regard to one of these three beliefs (which is, again, randomly chosen). Note that this simple model of a doxastic dynamic mimics the process of presenting internal paradoxes to a proponent; it does, however, not keep track of the paradoxes put forward so far, allowing the proponent (a) to fall back into a position previously occupied and thence (b) to oscillate between different positions forever. Yet this merely means that the simple model contains even less mechanisms which might compel the proponent to approach the truth than our debate simulations. Still, even in the simple model, proponents do eventually approach the truth, as Fig. 13.7 demonstrates. So the simple fact that proponents never face internal inconsistencies which comprise but correct individual sentences is sufficient to explain why rendering positions dialectically incoherent drives proponents, in the long run, closer to the truth.

Chapter 14
The Veritistic Dynamics of Argumentation Strategies in Many-Proponent Debates

We have found, in the previous chapter, that the argumentation strategies pursued in two-proponent debates have a significant influence on the veritistic dynamics. In this chapter, we will investigate whether this holds for many-proponent debates as well. In order to do so, we take the two most truth-conducive argumentation rules studied so far—*convert* and *undercut*—and modify them with a view to many-proponent debates. Specifically, the modified *convert* rule (*t-multiple convert*) tells proponents to introduce an argument whose premisses are shared by as many opponents as possible. Moreover, the *t-multiple convert* strategy prescribes that an opponent position be *rendered incoherent* once a full consensus has emerged. As we will see, this amounts to a crucial modification of the simple *convert* rule, where an argument that effectively *fortifies* the consensus is introduced provided that all proponents fully agree. Likewise, the modified *undercut* rule stipulates that a proponent undercuts, and thence renders incoherent, as many opponent positions as possible. As in the case of *t-multiple convert*, the *t-multiple undercut* rule also entails that a full consensus position, reached by all proponents, be rendered incoherent (if possible).

We will study how these argumentation strategies compare with a purely random argumentation investigated in Chap. 11 while paying particular attention to the veritistic dynamics, the significance of consensus as an indicator of truth, and the significance of stability as an indicator of truth.

14.1 Setup

For each argumentation strategy, we build an ensemble of 1,000 debates. The debates in the *t-multiple undercut* ensemble are set up as follows:

Argumentation Mechanism: Proponents, in alternating sequence, introduce arguments into the debate according to a modified *undercut* strategy. If (1) a full consensus has not yet been reached, a proponent, when it's her turn, first of

G. Betz, *Debate Dynamics: How Controversy Improves Our Beliefs*,
Synthese Library 357, DOI 10.1007/978-94-007-4599-5_14,

all identifies a sentence c she considers true while maximizing the number of opponents who don't agree with c. In other words, a proponent i at step t selects a sentence $c \in S$ such that (a) $\mathscr{P}_t^i(c) = \text{true}$ and (b) $|\{j|\mathscr{P}_t^j(c) = \text{false}\}|$ is maximal. In a second step, she determines all pairs of sentences (excluding $c/\neg c$) such that the number of opponents who accept both sentences yet disagree with c is maximal. Formally, these two distinct sentences $p_1, p_2 \in S \setminus \{c, \neg c\}$ maximize $|\{j|\mathscr{P}_t^j(c) = \text{false} \wedge \mathscr{P}_t^j(p_1) = \mathscr{P}_t^j(p_2) = \text{true}\}|$. The proponent then introduces an argument with conclusion c and one of these sentence pairs as premisses—taking into account the extra condition that adding this argument to τ leaves the true position coherent, that is, that the argument be valid. If, however, all proponents (2) have agreed on a full consensus position, this very consensus is undercut. We shall refer to this argumentation strategy as *t-multiple undercut*.

Discovery Mechanism: The background knowledge $\mathscr{B}$ is empty.

Update Mechanism: *Closest coherent* (cf. Sect. 4.1).

In the *t-multiple convert* ensemble, the debates' specification reads:

Argumentation Mechanism: Proponents, in alternating sequence, introduce arguments into the debate according to a modified *convert* strategy. If (1) not all proponents have reached a full consensus yet, the proponent i who may put forward the next argument chooses randomly, in a first step, a sentence c she considers true ($\mathscr{P}_t^i(c) = \text{true}$). In a second step, she determines all pairs of sentences (excluding $c/\neg c$) such that the number of opponents who accept both sentences is maximal. Technically, these two distinct sentences $p_1, p_2 \in S \setminus \{c, \neg c\}$ maximize $|\{j|\mathscr{P}_t^j(p_1) = \mathscr{P}_t^j(p_2) = \text{true}\}|$. The proponent then introduces an argument with conclusion c and one of the sentence pairs as premisses—taking into account the extra condition that adding this argument to τ leaves the true position coherent, that is, that the argument be valid. However, if (2) the proponents have already reached a full consensus position (which is not identical with the truth), the proponent introduces a valid argument that renders the consensus position dialectically incoherent.[1] We shall refer to this argumentation strategy as *t-multiple convert.*

Discovery Mechanism: The background knowledge $\mathscr{B}$ is empty.

Update Mechanism: *Closest coherent* (cf. Sect. 4.1).

The debates contain six proponents who pursue the ensemble's corresponding argumentation rule. The debate simulation terminates if all proponents have reached the truth or if a density greater than 0.8 is attained.

[1]Note, again, the major difference to the simple *convert* strategy, where the proponent introduces an argument which effectively fortifies the consensus position if there is no disagreement.

14.2 Results

14.2.1 *Truth's Attraction: How Rapidly Does the Proponents' Verisimilitude Increase?*

Figure 14.1 provides the basic information about the different debates' and argumentation strategies' veritistic dynamics. Unexpectedly, given the relative

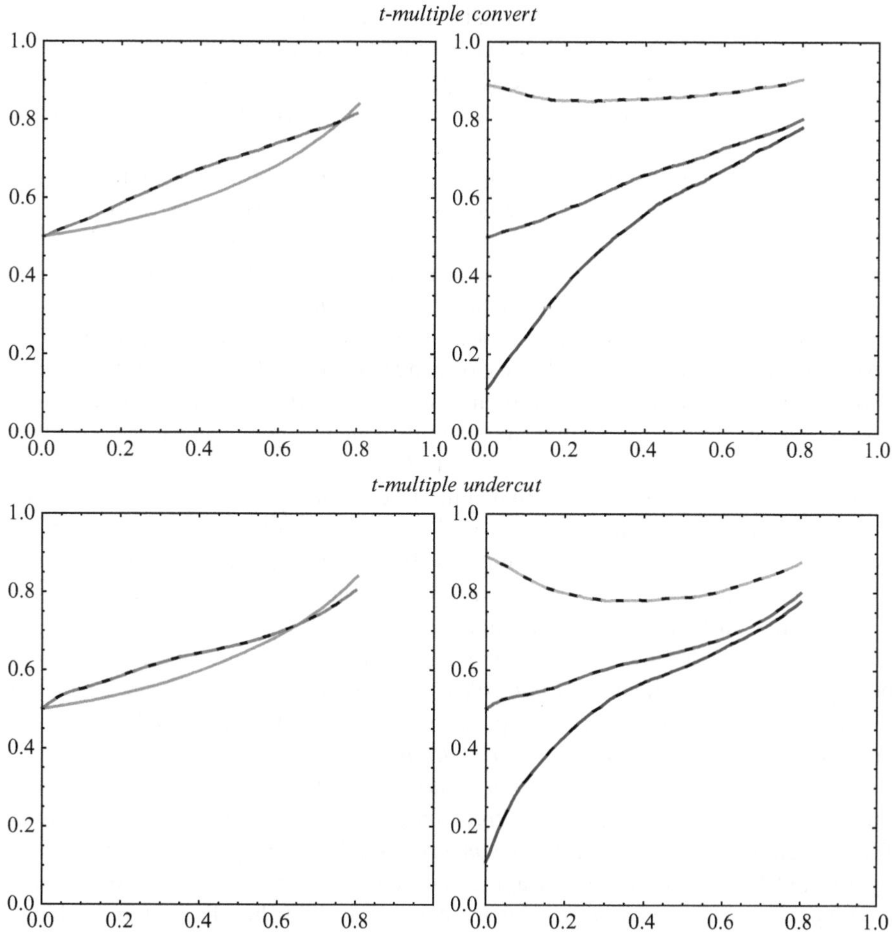

Fig. 14.1 Ensemble-wide mean verisimilitude evolutions, as functions of inferential density, averaged over all proponent positions (*left*) and proponents with specific initial agreement (*right*). The lower curve in the *left-hand* panels displays, as a point of reference, the mean verisimilitude evolution in the ensemble with *t-random argumentation* (see Fig. 11.1). The curves in the *right-hand* plots represent, from top to bottom, verisimilitude evolutions of proponents with high (0.8–1), medium (0.4–0.6), and low (0–0.2) initial verisimilitude

performance of *convert* and *undercut* assessed in the previous chapter, *t-multiple convert* is, in general, more truth-conducive than *t-multiple undercut*. At a density of $D = 0.5$, the proponents display, on average, a verisimilitude of 0.7 when following the *t-multiple convert* rule, as compared to a verisimilitude of roughly 0.65 when implementing the *t-multiple undercut* strategy. Both strategies, however, give rise to a more truth-conducive controversy, in particular at low densities, than a *t-random argumentation*. The right-hand plots in Fig. 14.1 detail the qualitative differences between the argumentation strategies. Thus, proponent positions which are initially close to the truth (top curves) retain, by and large, their high verisimilitude with *t-multiple convert* yet are pushed away from the truth substantially with *t-multiple undercut*. The loss of initially high verisimilitude coincides, however, with a superior gain of verisimilitude regarding initially completely false proponent positions (bottom curves). Yet, concerning proponent positions with medium initial verisimilitude (middle curves), *t-multiple convert* outperforms *t-multiple undercut* again.

As Fig. 14.2 demonstrates, both argumentation strategies fare equally well in terms of the number of entirely true proponent positions. At a density of $D = 0.5$, 1.5 proponents have acquired, on average, a completely true position. At $D = 0.8$, this holds for half of the proponents. Regarding the number of entirely correct proponent positions, both *t-multiple convert* and *t-multiple undercut* are much more truth-conducive than *t-random argumentation*. The right-hand plots in Fig. 14.2 reveal, notwithstanding the similar performance of *t-multiple convert* and *t-multiple undercut*, subtle differences. In particular, with *t-multiple convert*, proponents who display a high initial verisimilitude adopt a fully true position at an earlier stage than with *t-multiple undercut* (0.44 compared to 0.51). Vice versa, *t-multiple undercut* enables proponents with a very small initial verisimilitude to reach the truth at a lower density than *t-multiple convert* (0.81 compared to 0.86). These differences dovetail with the observed verisimilitude evolutions of proponents with fully correct or with completely false initial positions (cf. left-hand plots in Fig. 14.1).

We found, in Chap. 7, that the *multiple convert* strategy is substantially more *consensus-conducive* than *multiple undercut* and that this superior performance can be explained by the tendency of *multiple convert* to generate appropriate clusters in the debates' SCP where proponent positions are assembled and which are rapidly compressed. Can the performance of *t-multiple convert* be explained along the same lines? First of all, we may note that *t-multiple convert*, too, gives rise to much more fragmented debates than *t-multiple undercut*. As Table 14.1 details, the most fragmented debates in the *t-multiple convert* ensemble display a significantly lower aggregated NCC than the most fragmented debates in the other ensembles. Thus, *t-multiple convert*, like *multiple convert*, seems to generate highly clustered debates. But how does this relate to the veritistic dynamics?

The fragmentation of the SCP influences, as Fig. 14.3 shows, the veritistic dynamics of the debates. In particular, it seems to account for the superior performance of *t-multiple convert* since proponents in the most fragmented debates with *t-multiple convert* (dashed line in the upper left panel) display outstandingly high verisimilitude values. In these debates, 80% of the proponents' individual

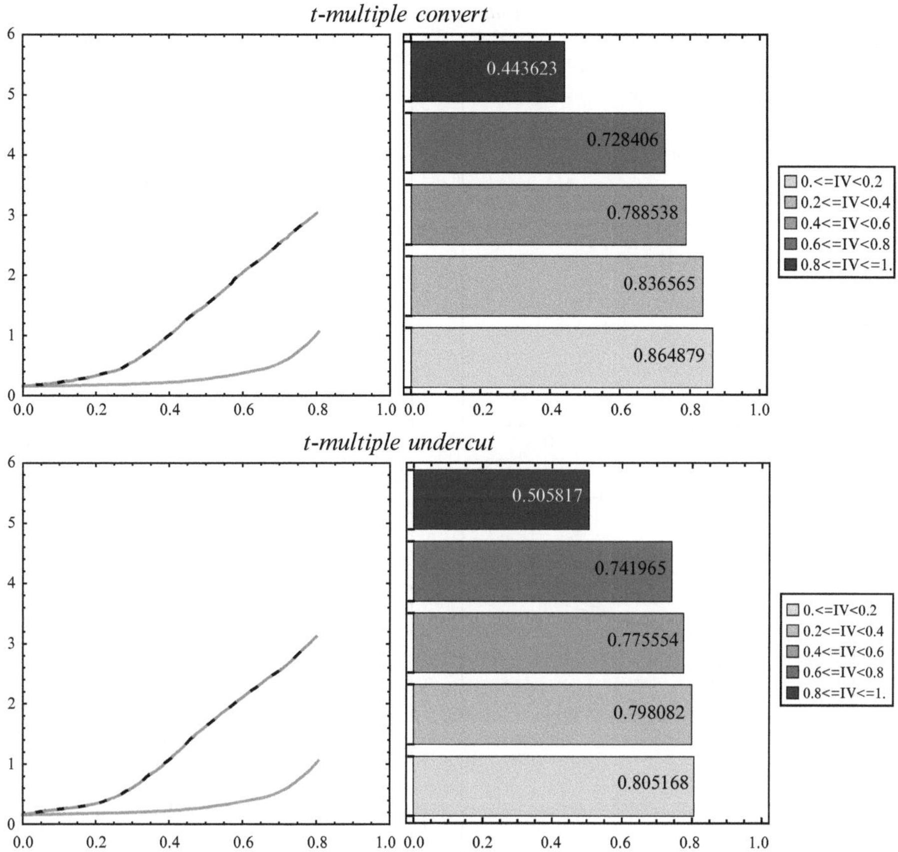

Fig. 14.2 Ensemble-wide average number of entirely true proponent positions as function of inferential density (*left*) and mean collapse-to-truth densities for proponent positions with specific initial agreement (*right*). To calculate the average collapse densities, we assume that proponents which haven't reached the truth when the simulation terminates ($D > 0.8$) acquire a fully correct position at $D = 1$. The lower gray curves in the left-hand plots reproduce, as a point of reference, the evolution of the number of entirely true proponent positions in the ensemble with *t-random argumentation* (See Fig. 11.2)

Table 14.1 Fragmentation of the SCP, measured by aggregated NCC, in different ensembles

Ensemble	Lower 10th quantile	Ensemble-wide mean	Upper 10th quantile
t-random argumentation	1.049	1.093	1.139
t-multiple undercut	1.009	1.0328	1.057
t-multiple convert	0.951	1.008	1.058

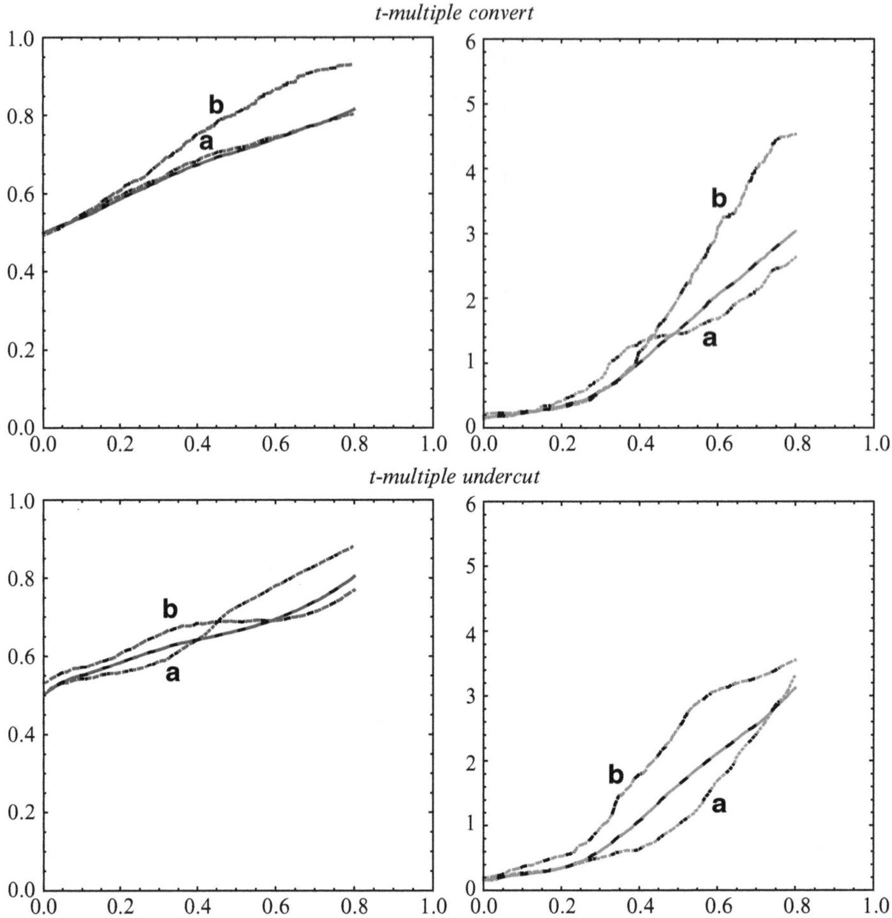

Fig. 14.3 Ensemble-wide mean verisimilitude evolutions (*left*) and average number of fully true proponent positions (*right*) in fragmented and compact debates, plotted as functions of inferential density. The different curves are calculated by taking into account: all debates (*solid*), very compact debates (*dotted*, *a*), highly fragmented debates (*dashed*, *b*)

beliefs are correct at a density of $D = 0.5$, and the verisimilitude reaches almost 95% at $D = 0.8$. The most compact debates, however, give rise to verisimilitude evolutions very similar to the ensemble mean. The fragmentation of the SCP plays a different role in the ensemble with *t-multiple undercut*. Here, proponents are, at low densities, closer to the truth in fragmented debates than in compact ones but possess, vice versa, a greater verisimilitude in compact debates than in fragmented ones at high densities. So the effect of fragmentation with *t-multiple undercut* is ambiguous.

Let us now turn to the number of entirely correct proponent positions and the way it is affected by fragmentation (right-hand plots in Fig. 14.3). In the long run,

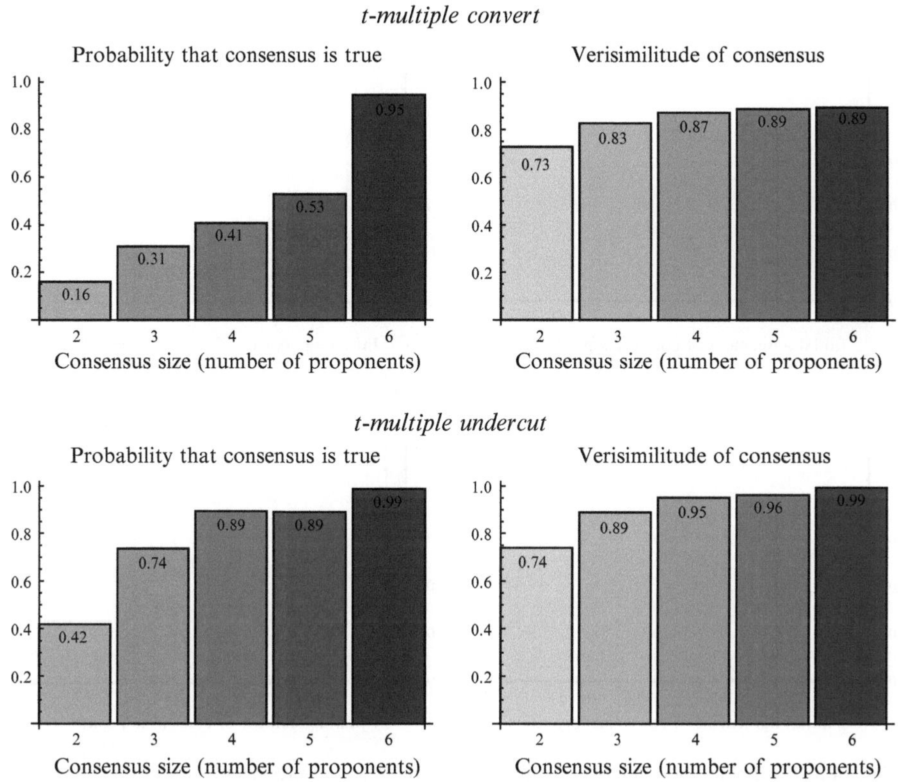

Fig. 14.4 Ensemble-wide relative frequency at which a consensus represents the true position as a function of the number of proponents who have come to agree (*left-hand* plots) and ensemble-wide mean verisimilitude of a consensus position as a function of the number of proponents who have come to agree (*right-hand* plots)

that is, at sufficiently high densities, fragmentation increases, in both ensembles, the number of fully true proponent positions per debate. At low densities, however, fragmentation is beneficial in the *t-multiple undercut* ensemble yet doesn't increase the number of proponents holding a fully true position with *t-multiple convert.*

14.2.2 The Verisimilitude of Consensus Positions: Is Mutual Agreement a Good Indicator of Having Reached the Truth?

We shall now study whether and under which conditions consensus is a reliable indicator of truth in debates with *t-multiple convert* and *t-multiple undercut.*

Figure 14.4 plots, for this chapter's ensembles, the likelihood of a consensus representing the truth, $\mathcal{T}$, and the consensus verisimilitude as a function of

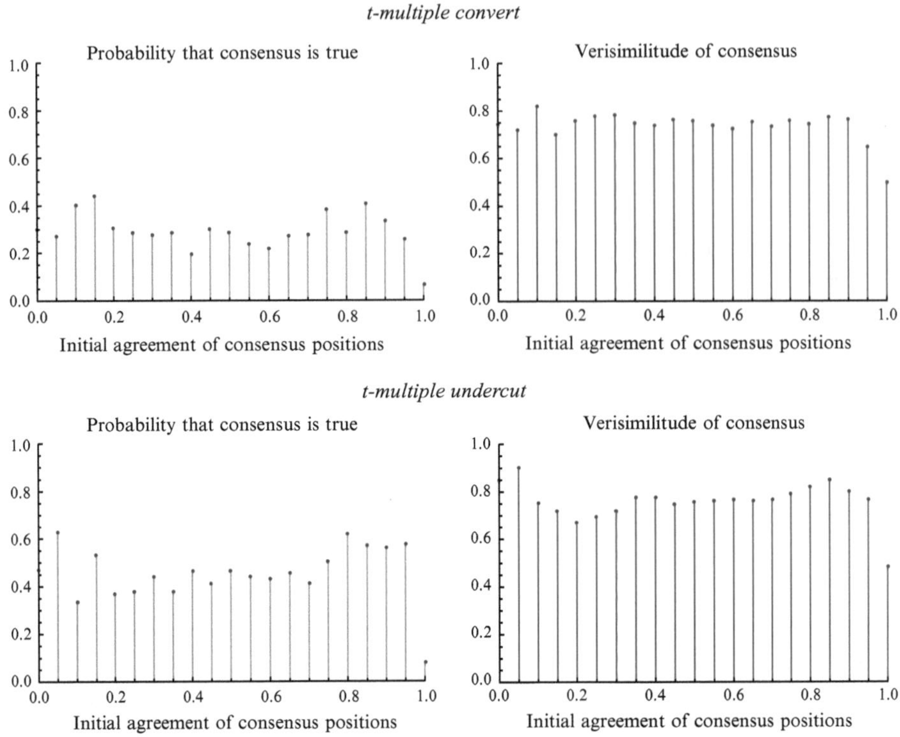

Fig. 14.5 Ensemble-wide relative frequency at which a two-proponent consensus represents the true position (*left-hand* charts) and ensemble-wide mean verisimilitude of a two-proponent consensus (*right-hand* charts), both plotted as a function of initial agreement between the two proponents who join the consensus

consensus size. With increasing size, consensus becomes generally more reliable as an indicator of truth. Comparing the two ensembles, we find that consensus is more revealing with *t-multiple undercut.* Here, for example, almost 3/4 of the consensus positions which are agreed upon by exactly three proponents are identical with the true position, as compared to 1/3 for *t-multiple convert* (and, similarly, 1/3 for *t-random argumentation*, see Fig. 11.6). Likewise, the verisimilitude of a consensus position which is reached with *t-multiple undercut* is typically greater than the verisimilitude of a corresponding consensus under *t-multiple convert* (and *t-random argumentation*).

In Chap. 11, we've seen that the higher the initial agreement between two proponents, the less telling (in terms of verisimilitude) is a consensus these proponents might eventually reach. But as Fig. 14.5 demonstrates, initial agreement does apparently not affect the verisimilitude of a consensus position in this chapter's ensembles. The verisimilitude and the likelihood that the consensus represents the truth stay roughly the same as the initial agreement between the consensus' members varies.

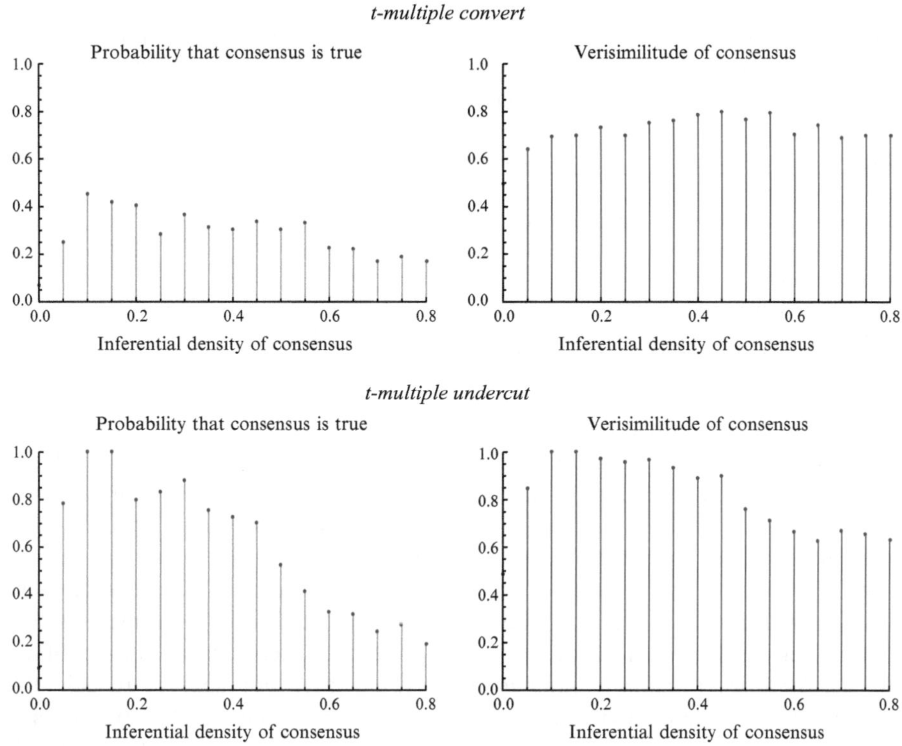

Fig. 14.6 Ensemble-wide relative frequency at which a 2-proponent-consensus represents the true position (*left*-hand charts), and ensemble-wide mean verisimilitude of a 2-proponent-consensus (*right*-hand charts), both plotted as a function of the inferential density at which the corresponding consensus emerges

Finally, does the accuracy of consensus as a veritistic indicator depend on the inferential density at which the corresponding consensus emerges? It does, as it turns out, but in precisely the opposite way as in the case of *t-random argumentation*. As the upper right panel of Fig. 14.6 demonstrates, there is no clear relationship between a two-proponent consensus' verisimilitude and its inferential density in the *t-multiple convert* ensemble. Yet the likelihood that a two-proponent consensus represents the truth seems to decrease slightly as the inferential density increases. This negative relationship is much more pronounced regarding *t-multiple undercut*. A two-proponent consensus which emerges at a low inferential density is very likely identical with the truth (>80%). At higher densities, however, the probability that a consensus among two proponents represents the truth declines substantially. Similarly, the verisimilitude of a two-proponent consensus is typically very high at low densities, and much smaller at higher densities. In sum, we have a highly accurate indicator of verisimilitude for very low densities with *t-multiple undercut*.

Table 14.2 Parameters of independent variables in a linear model that is fitted to the ensemble data. Each row displays the values corresponding to a linear model which explains that a consensus represents the truth, respectively its verisimilitude, in terms of the three independent variables

	Weights of independent variables		
Dependent variable	Normalized consensus size	Initial agreement	Inferential density
t-multiple convert			
Consensus is fully true (1) or not (0)	0.82	0.13	−0.13
Verisimilitude of consensus	0.26	0.12	0.07
t-multiple undercut			
Consensus is fully true (1) or not (0)	1.28	0.46	−0.64
Verisimilitude of consensus	0.52	0.22	−0.17

A multivariate regression analysis confirms, and generalizes, the observations which are based on the different graphs (cf. Table 14.2). Consider the *t-multiple convert* rule first. The likelihood that a given consensus represents the truth primarily hinges on the consensus size[2] (regression coefficient 0.82) and depends only marginally on the proponents' initial agreement (0.13) and the inferential density (−0.13). Concerning a consensus' verisimilitude, the proponent's initial agreement is comparatively more important, being half as influential (0.12) as consensus size (0.26). Yet, again, the inferential density doesn't affect a consensus' verisimilitude (0.07). As to the *t-multiple undercut* strategy, the regression analysis reveals that all three factors exert a significant influence on the probability that a consensus is identical with the true position. While consensus size is the dominant factor (1.28), both initial agreement (0.46) and inferential density (−0.64) have a notable—positive and negative, respectively—impact as well. This fairly significant and, puzzlingly, positive role of initial agreement is newly revealed by the regression analysis, and hasn't been apparent in Fig. 14.5. Regarding verisimilitude, consensus size is, again, the most influential factor (0.52), with initial agreement (0.22) and inferential density (−0.17) being, in absolute terms, less than half as important.

14.2.3 The Verisimilitude of Stable Positions: Are Proponent Positions Which Remain Relatively Stable Closer to the Truth?

In Chap. 11, exploring the idea that the stability of proponent positions is an indicator of the corresponding positions' verisimilitude, we have found that, in random debates, highly stable positions tend to be closer to the truth, indeed. This holds, we have seen, for stability defined as the frequency of previous falsifications

[2]We consider, more precisely, *normalized* consensus size so that all independent variables vary between 0 and 1.

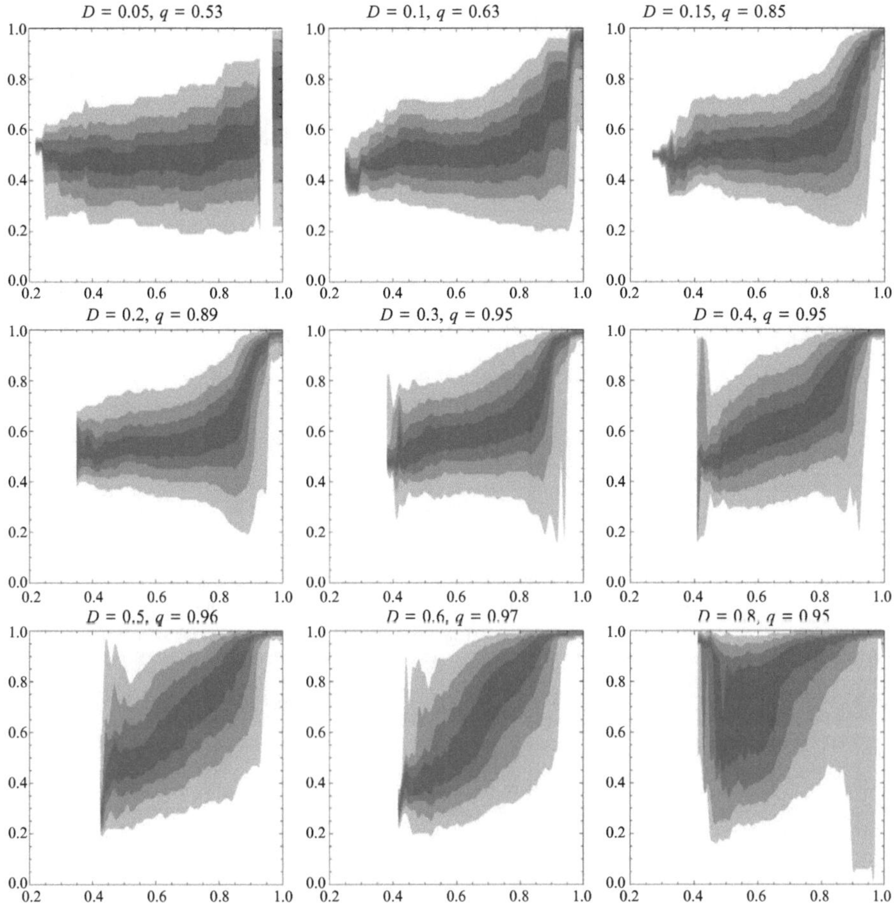

Fig. 14.7 Verisimilitude of proponent positions, at a certain density *D*, as a function of their stability—defined as the relative frequency at which the proponent's previous positions have been rendered incoherent—in the ensemble with *t-multiple convert* (see Fig. 11.10 for a detailed description)

as well as for stability as measured by the agreement with the corresponding initial position. We shall now investigate whether these results apply to debates with *t-multiple convert* and *t-multiple undercut* too.

In a first step, we consider stability as approximated by the relative frequency at which a proponent's previous positions have been rendered incoherent. As Figs. 14.7 and 14.8 demonstrate, we obtain a neat positive relationship between stability and verisimilitude in both ensembles. In particular, this relationship holds already at very low densities. The fraction of highly stable positions which are also close to the truth, q, allows us to quantify the strength of this relation. Thus, at a density of $D = 0.15$, 85% of the highly stable proponent positions are, in the *t-multiple convert*

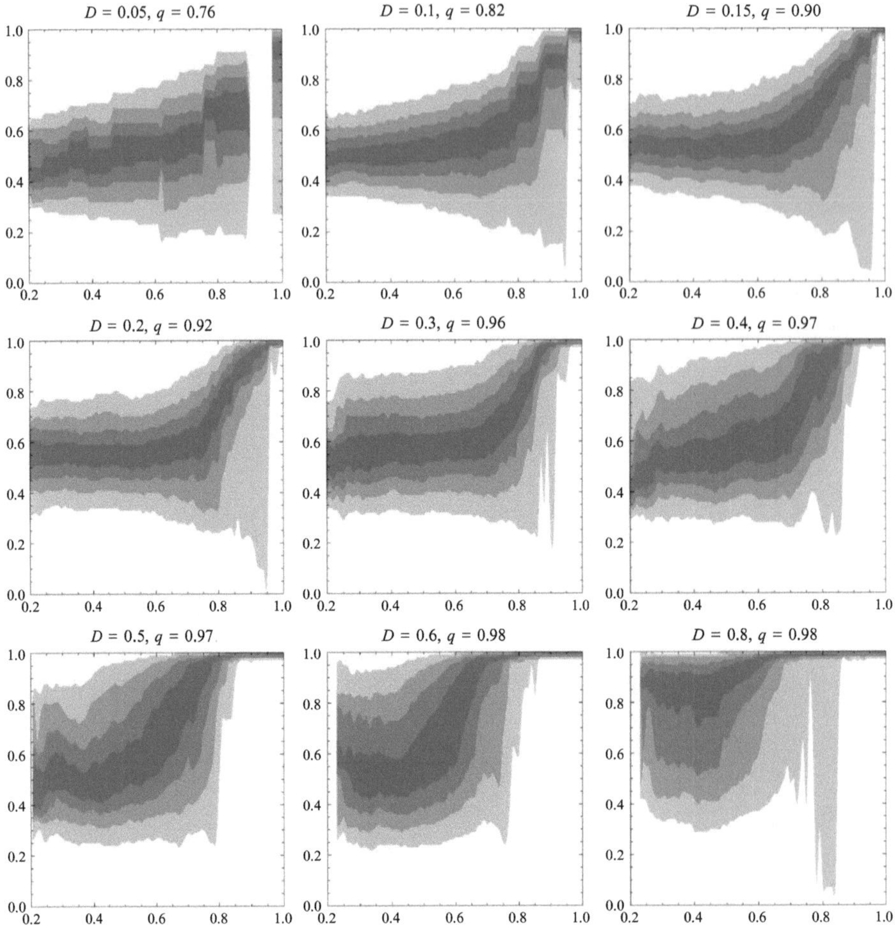

Fig. 14.8 Verisimilitude of proponent positions, at a certain density D, as a function of their stability—defined as the relative frequency at which the proponent's previous positions have been rendered incoherent—in the ensemble with *t-multiple undercut* (see Fig. 11.10 for a detailed description)

ensemble, close to the truth (verisimilitude greater than 0.8).[3] This compares with merely 30% in random debates (cf. Fig. 11.10). In the *t-multiple undercut* ensemble even 90% of highly stable proponent positions display a verisimilitude greater than 0.8 at $D = 0.15$. Hence, the relative frequency of falsifications, and simultaneous readjustments, yields a most accurate indicator of verisimilitude in these ensembles even at very low densities.

[3]Note that, at $D = 0$, a fraction of 16% of the highly stable (= all) proponent positions is close to the truth.

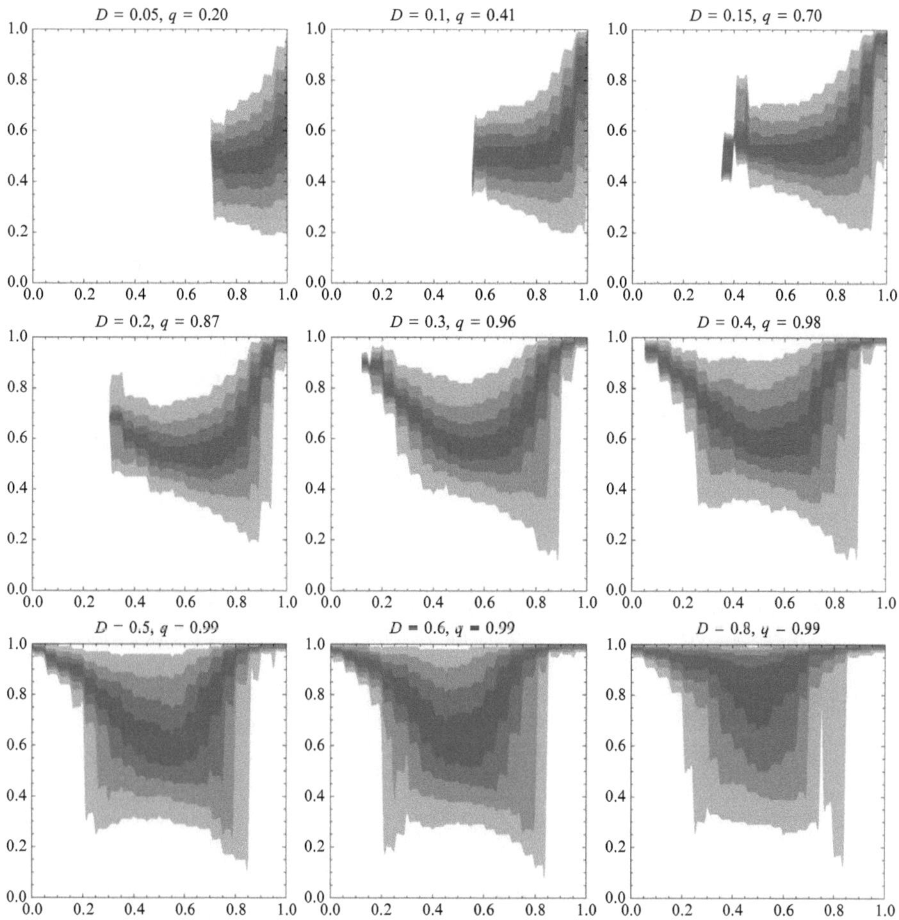

Fig. 14.9 Verisimilitude of proponent positions, at a certain density D, as a function of their stability—approximated by their distance to the corresponding initial position—in the ensemble with *t-multiple convert* (see also Fig. 11.11)

Let us now turn to the second approximation of stability, namely the agreement of a proponent position with the proponent's initial position. As Figs. 14.9 and 14.10 show, the relationship between agreement with one's initial position and verisimilitude is intricate. At low densities, we discern a positive dependence similar to the one observed in Figs. 14.7 and 14.8. At higher densities, however, a U-shaped relationship gradually emerges. Accordingly, proponent positions which exhibit extremely low or very high agreement with the corresponding initial position tend to be close to the truth; positions with medium stability, in contrast, tend to be rather remote from the truth. This observation fits well with the results concerning *t-random argumentation*. Again, the fraction of highly stable positions which are also close to the truth helps us to quantify the strength of the observed relationship.

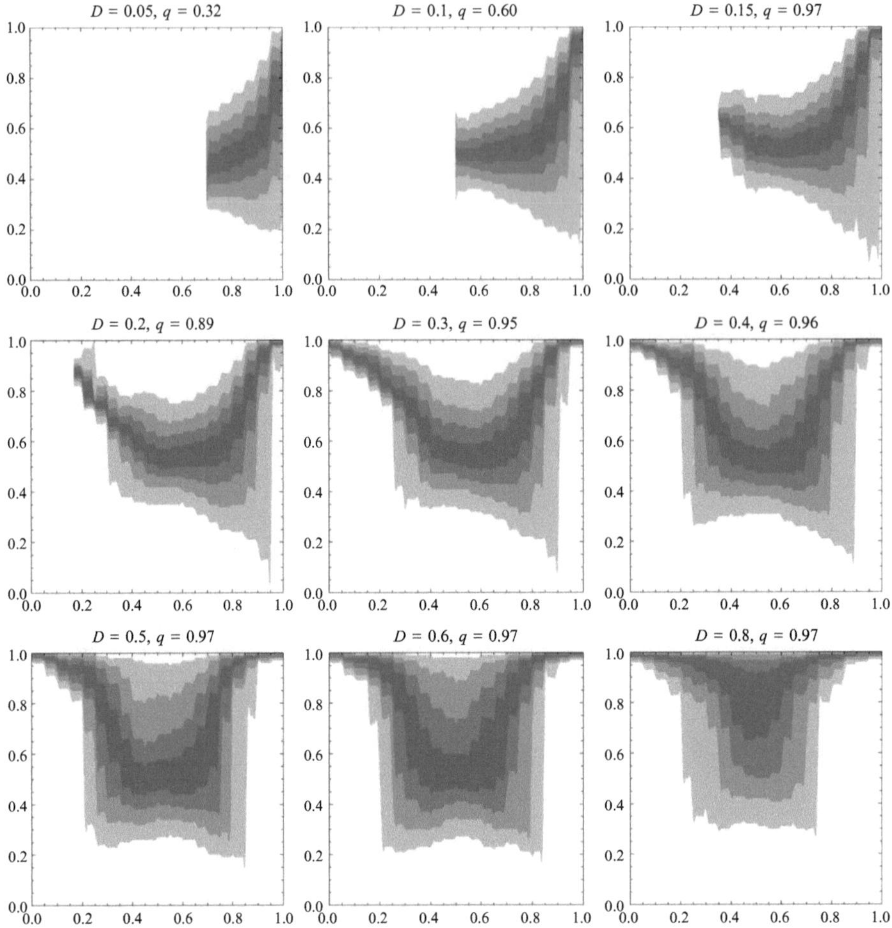

Fig. 14.10 Verisimilitude of proponent positions, at a certain density D, as a function of their stability—approximated by their distance to the corresponding initial position—in the ensemble with *t-multiple undercut* (see also Fig. 11.11)

In the ensemble with *t-multiple convert*, 70% of highly stable positions display a verisimilitude greater than 0.8 at a density of $D = 0.15$. This is substantially higher than the corresponding fraction of 22% we obtained in random debates (see Fig. 11.11). Yet, with *t-multiple undercut*, even 97%—almost *all*—of the stable proponent positions are close to the truth. Employing the *t-multiple undercut* strategy increases the accuracy of stability as an indicator of truth substantially (both in the sense of frequency of falsifications and in the sense of agreement with one's initial position), thereby enabling us, in principle, to make reliable inferences about the verisimilitude of proponent positions at low densities.

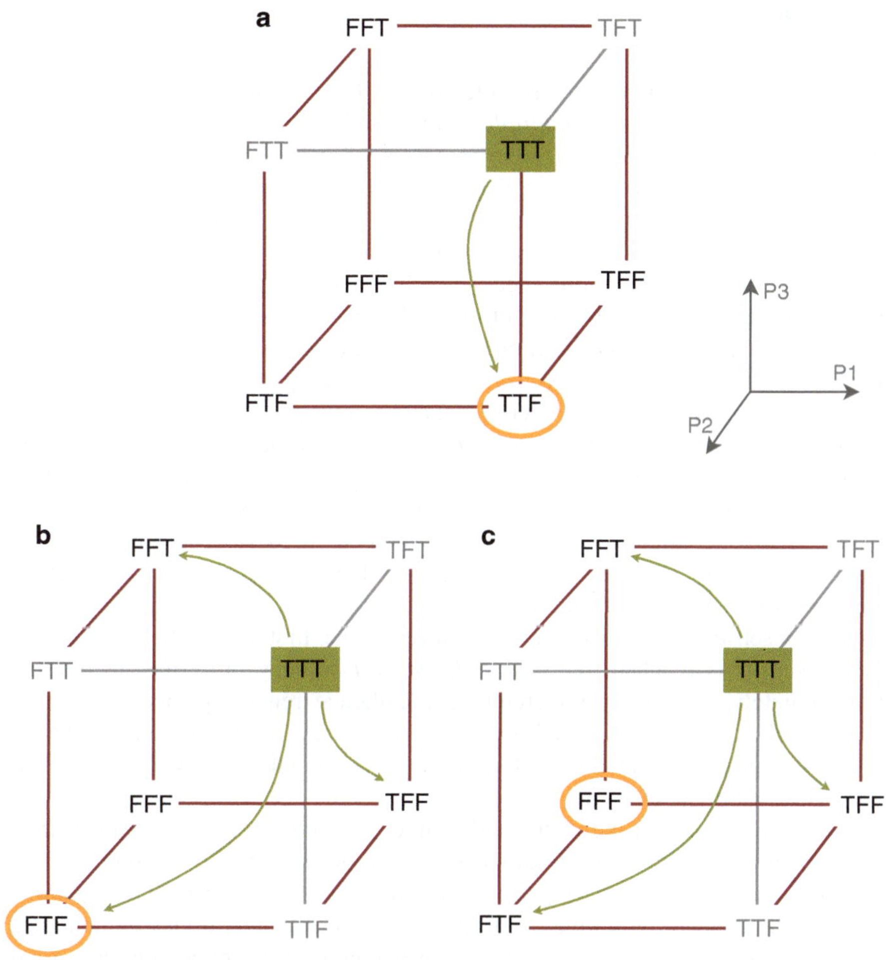

Fig. 14.11 Closest coherent positions of a maximally fortified, spurious consensus in an illustrative SCP (see Fig. 6.5 for more details on this kind of visualization). As the consensus position (petrol) is maximally fortified, every neighboring position, apart from the truth (*orange circle*), is rendered incoherent (*gray coloring*). Provided the consensus position is itself undercut, proponents are compelled to resort to one of the closest coherent positions as indicated by the petrol arrows. In case the verisimilitude of the spurious consensus is greater than 0.5 (**a**), proponents will directly adopt the truth. In case the verisimilitude of the spurious consensus is less than 0.5 (**b** and **c**), the proponents are bound to adopt positions which are in any case not more remote from the truth than the spurious consensus itself and possibly much closer. This increases the mean verisimilitude, on average, from 3/9 to 5/9 (**b**) and from 0 to 2/3 (**c**)

14.3 Discussion

The following discussion of the reported results focuses on three points. First, we will try to understand why *t-multiple convert*, unlike the simple *convert* strategy in dualistic debates, is so effective in terms of increasing truthlikeness, even outperforming *t-multiple undercut*. Explaining this observation is the main purpose of our discussion. In a much more concise way, we will try to understand, secondly, why stability represents a highly accurate indicator of truth with *t-multiple undercut*. Finally, we consider the more specific finding that consensus becomes, *ceteris paribus*, more telling an indicator of truth with lower inferential density.

To start with, recall that the limited truth-conduciveness of the simple *convert* rule stems primarily from the *convert* rule's outstanding consensus-conduciveness and the ensuing tendency to generate *spurious* consensus. So, to understand the superior truth-conduciveness of *t-multiple convert*, we have to comprehend why proponents in the ensemble with *t-multiple convert* are less prone to get caught in a spurious consensus. There are two obvious answers. First, the debates studied in this chapter contain six proponents as compared to two proponents in the dualistic controversies studied in the previous chapter. Consequently, the debates with *t-multiple convert* are characterized by a greater diversity of opinion, which makes consensus more difficult to achieve in the first place. Second, once a full consensus is reached, the *t-multiple convert* strategy prescribes to question it immediately, effectively undercutting and thence rendering incoherent every spurious full consensus. So even if *t-multiple convert* causes the proponents to agree on a false consensus, they are not locked in (by continually fortifying the spurious consensus) but break out of it.

Let us, next, consider the veritistic dynamics of *t-multiple convert* and *t-multiple undercut* in more detail. The fragmentation of a debate's SCP represents the key to understanding the superior performance of *t-multiple convert*. The *t-multiple convert* rule does not only give rise to debates with unusually high fragmentation; these highly fragmented debates are, moreover, characterized by a rapid increase of verisimilitude, as Fig. 14.3 demonstrates. The following qualitative description of the position dynamics in these debates may help to understand this fact and provide an explanation: Each extremely fragmented debate begins with a phase in which the proponents succeed in shaping the SCP and in forcing each other to readjust their positions so that they gather on one and the same, relatively isolated cluster of the SCP. By gradually shrinking this cluster, they approach each other quickly (see Chap. 7). If, coincidentally, the true position belongs to that very cluster, the proponents will quickly find it. Yet if it doesn't, then the rapid agreement within the shrinking cluster of the SCP leads to spurious consensus. Because of the possibility of such a spurious consensus, highly fragmented debates don't display superior performance in terms of truth-conduciveness at low densities (see the dashed lines in Fig. 14.3). Now, assume that the proponents have come to agree on a spurious consensus. As noted above, this incites, qua definition of the argumentation rule, an immediate falsification of that very consensus position, causing all proponents to adjust their positions and pushing them *simultaneously* towards the truth. It is because of these collective falsifications that the verisimilitude as well as the number of proponents with fully correct proponent positions starts to increase rapidly at medium densities. Apparently, it is much more efficient (1) to assemble different proponents (which are all possibly quite remote from the truth) in a first step and then to push them, collectively, towards the truth in a second step, than (2) to push different individual positions with significant mutual disagreement towards the truth independently of each other. By creating a possibly spurious consensus in a first phase of a debate, the proponents are, in the second phase, able to address all proponent positions simultaneously with each new argument introduced.

To conclude this explanation, we have to understand why rendering a spurious consensus incoherent tends to improve the verisimilitude of the proponent positions rather than to push the proponents (simultaneously) further away from the truth. Why, in other words, is a spurious consensus typically corrected in the right way? We have discussed a similar question at the end of Sect. 13.3, where we explained the superior truth-conduciveness of critical argumentation in general: Arguments which render positions incoherent tend to target incorrect rather than correct beliefs. In regard to the correction of a spurious consensus, we may substantiate and detail this explanation as follows. As the isolated cluster in the SCP on which the proponents have gathered is gradually contracting, more and more positions in that very part of the SCP are rendered incoherent. If a spurious consensus eventually emerges on an isolated cluster, the consensus position is effectively fortified, that is, surrounded by incoherent positions. Let us assume, for the sake of illustration, that virtually the entire neighborhood of such a spurious consensus be rendered incoherent. Why does a falsification of the consensus position increase, on average, the proponents' verisimilitude? Suppose, case one, the true position is close to the spurious consensus. That implies that the truth represents the only position in the vicinity of the consensus which is not rendered incoherent. Therefore, all proponents will adopt the truth, for being the closest coherent position, as soon as the consensus is rendered incoherent (see Fig. 14.11a). If, case two, the spurious consensus possesses an extremely low verisimilitude, then all similarly incorrect positions are rendered incoherent, and the proponents' new positions will necessarily be closer to the truth (cf. Fig. 14.11c). If, finally, the consensus position displays medium verisimilitude, its falsification will cause the proponents, at least under ideal assumptions and on average, to increase their verisimilitude too, as illustrated in Fig. 14.11b.

The effects of fragmentation allow us to make sense of the particular qualitative dynamics of highly fragmented debates in the *t-multiple undercut* ensemble, too. As we found in Chap. 7, proponent positions are typically distributed on different clusters in those debates. Our metaphor of the flooded village neatly applies. The inhabitants are driven upon the roofs of different buildings by the rising flood. Some of them might coincidentally be pushed upon the castle's roof, close to its tower which cannot be flooded (the true position). The verisimilitude of these lucky few will increase rapidly, causing the increase of ensemble-wide mean verisimilitude and mean number of completely true proponent positions at low densities (dashed curves in Fig. 14.3). However, as the clusters are contracting, the proponents which are, by chance, located on the cluster with the true position will eventually find the truth, and their verisimilitude won't increase any further. The proponents, however, who are located on other clusters will be caught on isolated opinion islands, and be forced into spurious consensus. This effect causes the verisimilitude evolution as well as the evolution of the number of fully correct proponent positions to flatten at medium densities. So, the dynamic geometry of the SCP provides, once more, important insights which help to understand the veritistic dynamics of particular debates.

We have found, in the previous section, that *t-multiple undercut*, both as compared to *t-multiple convert* and as compared to *t-random argumentation*, improves the accuracy of stability as an indicator of truth substantially. This important finding can be explained straightforwardly. Recall that *t-multiple undercut* amounts to the most aggressive argumentation strategy we have designed so far. It prescribes to falsify as many opponent positions as possible. As a consequence, the rate of falsifications is much higher with *t-multiple undercut* than with *t-multiple convert* or *t-random argumentation*, and *t-multiple undercut* succeeds in rendering, with a given number of arguments, much more proponent positions (which are not entirely correct) incoherent than the other strategies. But this means that those proponent positions which remain relatively stable under *t-multiple undercut* contain a higher fraction of positions which simply cannot (easily) be rendered incoherent since they are true (or at least close to the truth). With *t-random argumentation* or *t-multiple convert*, in contrast, a much larger fraction of stable proponent positions is simply stable by chance, that is, because the corresponding positions have not been severely challenged. The increased aggressiveness of the argumentation is the reason why stability becomes such an accurate indicator of truth with *t-multiple undercut*.

In the remainder of this section, we shall briefly turn to the observation that consensus is, in the *t-multiple undercut* ensemble, a most accurate indicator of truth at low densities. This is, at first glance, surprising since it contrasts with our results regarding random debates. The finding, however, becomes intelligible if we consider the consensus-conduciveness of the *t-multiple undercut* strategy. *t-Multiple undercut*, like *multiple undercut*, represents a highly offensive and critical argumentation rule which excels particularly in destroying coincidental agreement and consensus. As the top curve in the bottom, right-hand plot in Fig. 7.1 testifies, contingent agreement is systematically knocked down at low densities. Only at higher densities, when the SCP is sufficiently fragmented and proponent positions are assembled on clusters, does *t-multiple undercut* engineer full agreement and thus generate, possibly, spurious consensus. Now, if an argumentation strategy effectively destroys full agreement (instead of engineering it), the consensus positions which prevail nonetheless can apparently not be abolished. As the only kind of consensus which cannot be abolished by *t-multiple undercut* (or any other strategy) is a fully correct consensus (as already noted above), a substantial proportion of the consensus positions at low densities represent the truth. At higher densities, however, when *t-multiple undercut* does engineer agreement, a greater fraction of the consensus positions attained are spurious.

With consensus on the one hand and stability on the other hand, we have found indicators of truth which are highly accurate at low densities, given the right argumentation strategy. This constitutes a relevant result with practical significance. It does not only allow us to accurately estimate the proponent positions' verisimilitude at low densities. It also accounts for the epistemic virtue of criticism and the importance of controversy in any epistemic enterprise: Engaging in a critical argumentation is a precondition for drawing this kind of inferences about the verisimilitude of proponent positions in a reliable way.

Chapter 15
The Veritistic Dynamics of Debates with Core Updating

In this chapter, we drop the assumption that the proponents in a debate deem all sentences equally important. More specifically, we presume, in analogy to Chap. 8, that a subset of the sentence pool contains the debate's core theses. Proponents are, accordingly, particularly reluctant to modify their convictions regarding these central claims and prefer, rather, to adjust the truth value assignments vis-à-vis the auxiliary sentences outside the debate's core. The introduction of core beliefs allows us to consider the robustness of the proponents' partial positions (i.e., their degree of justification), and we can investigate its bearing on the veritistic dynamics—which is one of the main purposes of this chapter. Whereas we study, in this chapter, the debate simulations with *t-random argumentation* and *lexicographic closest coherent* update mechanism, we will explore, in the ensuing, final chapter, the effect of argumentation strategies which take the distinction between core beliefs and auxiliary beliefs explicitly into account.

15.1 Setup

This chapter's ensemble comprises 1,000 debate simulations which are set up as follows:

Argumentation Mechanism: *t-Random argumentation* (cf. Sect. 11.1).
Discovery Mechanism: The background knowledge remains empty.
Update Mechanism: *Lexicographic closest coherent* (cf. Sect. 8.1).

Each debate contains six proponents. Their core positions are defined on a fixed set of five sentences. A debate simulation terminates as soon as all proponents have acquired a fully correct position, regarding both their core as well as their auxiliary beliefs.

G. Betz, *Debate Dynamics: How Controversy Improves Our Beliefs*, Synthese Library 357, DOI 10.1007/978-94-007-4599-5_15,

15.2 Results

15.2.1 Core Truth-Conduciveness

As Fig. 15.1a shows, the *lexicographic* update rule, or, to put it more generally, the proponents' reluctance to modify their core beliefs, decelerates the verisimilitude growth of the proponents' core positions markedly. In fact, the proponent core positions hardly approach the truth at all for densities lower than 0.5. The corresponding partial positions in debates with *t-random argumentation* and simple *closest coherent* update (light curve), in contrast, display a considerable verisimilitude increase in this very interval.

Similarly, the *lexicographic* update mechanism delays the rise of the number of fully correct core proponent positions, too (documented by Fig. 15.1b). Again, up to a density of $D = 0.5$, the number of proponents who hold entirely correct core positions hardly changes with *lexicographic* update. Between densities of 0.2 and 0.9, there are, on average, more proponents who hold a fully correct core position in the ensemble with the simple *closest coherent* update mechanism than in the ensemble with the *lexicographic* update rule.

The distinction between core and auxiliary beliefs enables us to calculate the degree of justification of the proponents' core positions. As in Chap. 8, we consider, for each proponent, the degree of justification of her core position at an early stage of

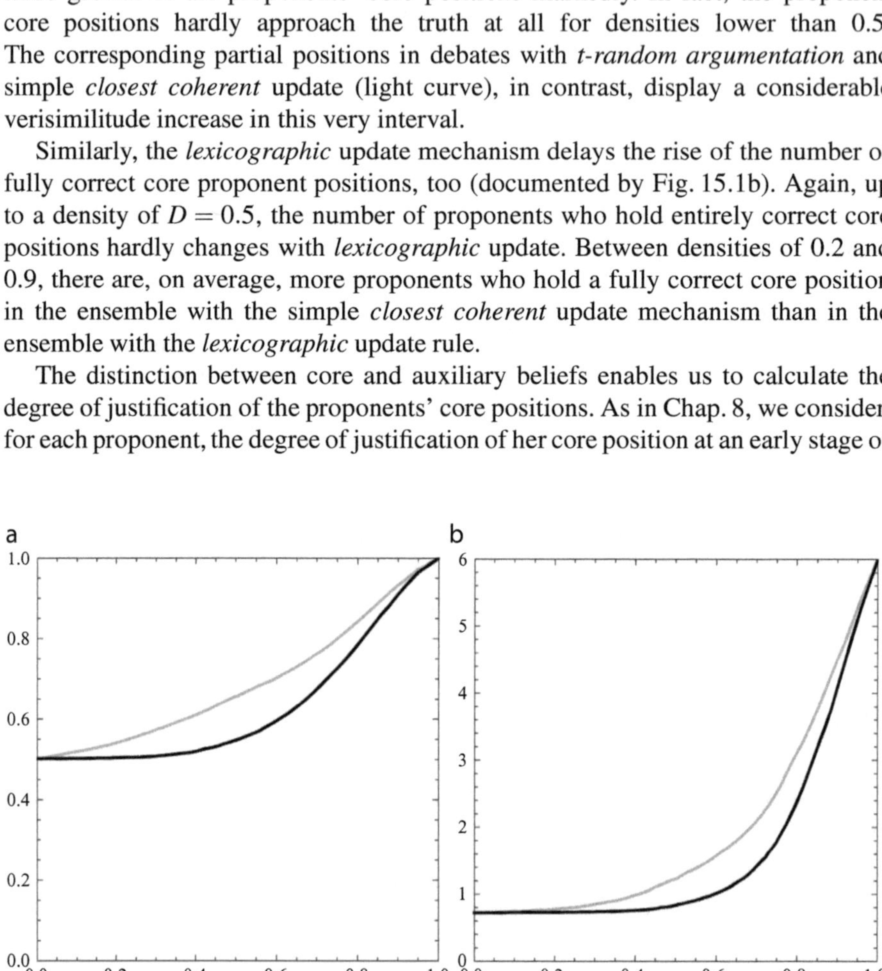

Fig. 15.1 Ensemble-wide mean verisimilitude of proponent core positions (**a**) and ensemble-wide mean number of fully correct proponent core positions (**b**) as functions of inferential density. Agreement evolutions are plotted for this chapter's ensemble with *lexicographic closest coherent* update (*dark curves*) and for the ensemble, presented in Chap. 11, with simple *closest coherent* (*light curves*). As regards the second case, the proponents' core beliefs are presumed to relate to the very same five sentences which make up the proponents' cores in the first ensemble. Note, however, that these core beliefs don't influence the corresponding debate dynamics because proponents update according to the simple *closest coherent* method and arguments are introduced randomly

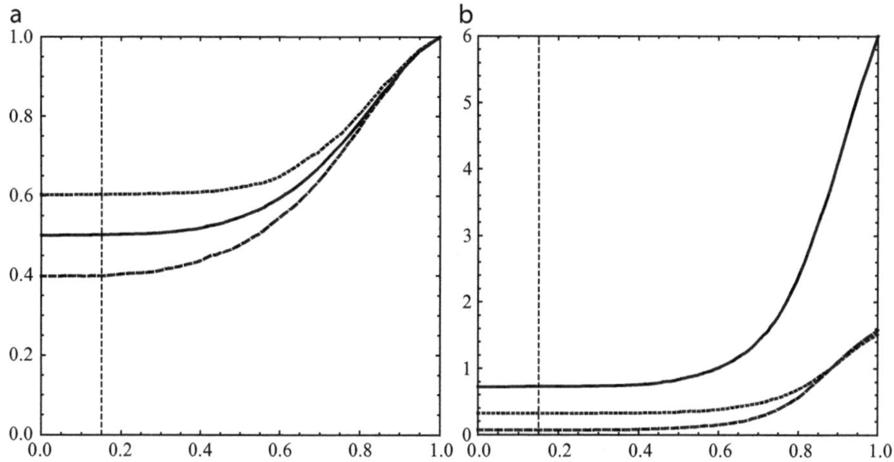

Fig. 15.2 Ensemble-wide mean normalized agreement of proponent core positions (**a**) and ensemble-wide mean number of fully true core positions (**b**)—all plotted as functions of inferential density. The plots display ensemble-wide means as averaged over all proponents (*solid curves*), proponents with a very robust core position at $D = 0.15$ (*dotted curves*), and proponents who hold a core position with very low robustness at $D = 0.15$ (*dashed curves*). More specifically, a partial core position with high (low) robustness possesses a degree of justification which falls in the upper (lower) quartile of all robustness scores at the corresponding density in the ensemble

the debate, more precisely, at a density of $D = 0.15$. Note that, at this density, mean verisimilitude and mean number of fully true core positions have hardly changed as compared to their initial values. Figure 15.2 juxtaposes the evolution of mean verisimilitude and number of fully correct core positions as regards all proponents (solid curves), proponents who hold a very robust position at $D = 0.15$ (dotted curves), and proponents who hold a position with a low degree of justification at $D = 0.15$. As Fig. 15.2a demonstrates, proponents whose core position possesses a high (low) degree of justification at $D = 0.15$ display in general an above-average (below-average) verisimilitude throughout the entire debate, including the initial state. Thus, proponent core positions with high (low) robustness at $D = 0.15$ are typically 10 percentage points closer (more distant) to the truth than the ensemble-wide average proponent core position. These differences gradually shrink as the inferential density approaches 1. The degree of justification of a core position has obviously an impact on the number of fully correct core positions as well. Figure 15.2b shows that there are, in absolute terms, more proponents with a robust and fully correct core than proponents with an unrobust, yet fully correct core position. In relative terms, this translates into a picture similar to Fig. 15.2a.[1] Core positions with a high (low) degree of justification contain an above-average (below-average) share of fully correct positions.

---

[1] As the dotted and dashed lines in Fig. 15.2b consider but a quarter of all proponents, one has to scale them by a factor of 4 in order to normalize them relative to the solid curve.

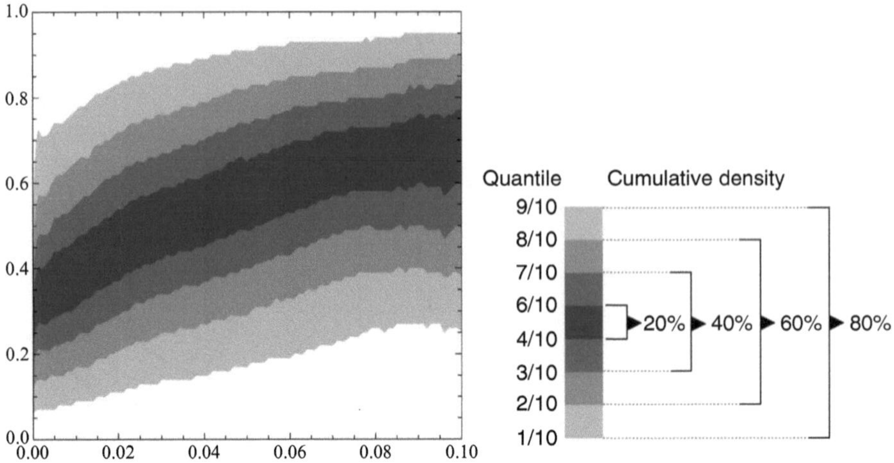

Fig. 15.3 Verisimilitude of proponent core positions as a function of their robustness at density $D = 0.15$. The diagram indicates, in relative terms, how many core positions with a certain robustness (x-axis) possess the corresponding verisimilitude (y-axis). The shading levels represent different quantiles as specified in the legend. The quantiles are calculated as follows: For each robustness value, a smooth probability density function (PDF) is fitted to the discrete relative frequencies of the corresponding verisimilitude values. This interpolated PDF is then used to derive the quantiles. In order to increase its accuracy, this plot is based on an ensemble of 5,000 instead of 1,000 debate simulations

15.2.2 Robustness of Proponent Core Positions and Verisimilitude

Figure 15.2 suggests that core positions with a high robustness at $D = 0.15$ tend to be closer to the truth. Figure 15.3 scrutinizes this hypothesis by directly plotting the relationship between a core position's degree of justification at $D = 0.15$ and its verisimilitude at that very inferential density. This reveals a clear-cut positive correlation. Thus, roughly half of the positions with very low degree of justification (<0.005) disagree with the truth by more than 70%. Likewise, highly robust core positions tend to be close to the truth. More specifically, a statistical analysis of the ensemble shows that 60% of the proponent core positions with a degree of justification greater than 0.09 display a verisimilitude of at least 0.8. In sum, the degree of justification, besides a positions' stability, turns out to be an accurate indicator of a proponent position's verisimilitude at low inferential densities.

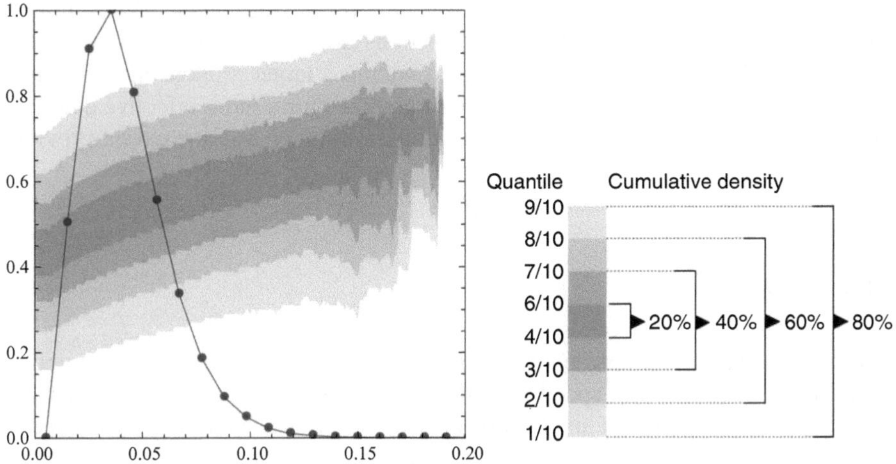

Fig. 15.4 Verisimilitude of all core positions as a function of their robustness at the density $D = 0.15$. In contrast to Fig. 15.3, this plot considers all core positions, no matter whether they are maintained by some proponent or not. The *shading* indicates, in relative terms, how many core positions with a certain robustness (x-axis) possess the corresponding verisimilitude (y-axis). The *circles* represent the frequencies at which core positions possess a degree of justification in the corresponding interval (of width 0.01). These frequencies are plotted relative to the maximum frequency in the sample. In order to increase its accuracy, this plot is based on an ensemble of 5,000 instead of 1,000 debate simulations (for further details, see Fig. 15.3)

15.2.3 General Correlation Between Degree of Justification and Verisimilitude

So far, we have considered the robustness of a proponent's core position as an indicator of her core position's verisimilitude—in close analogy to the way we have regarded a position's stability as a truth indicator in previous chapters. However, the core positions which are not held by any proponent whatsoever do, of course, possess a specific degree of justification, too, and the question arises whether their degree of justification is correlated with verisimilitude as well. Figure 15.4 plots the verisimilitude of all core positions, no matter whether they are actually maintained by some proponent or not, as a function of their degree of justification at $D = 0.15$. Taking into account all core positions significantly broadens the sample, which now includes core positions with more extreme robustness scores. Thence, Fig. 15.4 may cover a wider range than Fig. 15.3. The main result states, first of all, that degree of justification is positively correlated with verisimilitude as regards all core positions in a debate. In particular, more than half of the core positions with a degree of justification greater than 0.15 agree with the truth by more than 70%. But, as the histogram included in Fig. 15.4 shows, only a tiny fraction of all core positions exhibit degrees of justification of this size. Secondly, the relationship appears to be slightly less pronounced than in the case of proponent core positions depicted in

Fig. 15.3. Yet this last observation might simply result from the specific sampling procedure by which initial proponent positions are constructed: That procedure ensures that the set of proponent positions contains a larger proportion of positions with extreme verisimilitudes, and this in turn allows for a more effective detection of these positions by using robustness as an indicator (see also the caveat provided on page 176).

15.3 Discussion

The previous section has presented two main results. First, the *lexicographic* update mechanism decreases the speed at which proponent core positions approach the truth. Second, the degree of justification of a core position at a low density represents a significant indicator of verisimilitude. Now, the explanation of the first of these two results is straightforward. With *lexicographic*—as compared to simple—*closest coherent* update, proponents are much less prepared to modify their core beliefs and prefer to alter auxiliary convictions. As a consequence, core positions are changed less frequently, which necessarily delays their getting closer to the truth.

In the remainder of this chapter, we will discuss, and try to understand, the second main result, namely, that a core position's verisimilitude is positively correlated with its degree of justification. This result signifies that the arguments, which are introduced randomly, tend to increase the degree of justification of predominantly correct core positions to a larger extent than the degree of justification of mainly false core positions. Or, the arguments decrease the degree of justification of mainly incorrect positions to a larger extent. But for what reason? Let us consider all arguments which may lower or raise the degree of justification of a proponent's core position. More specifically, we consider all two-premiss arguments which contain one or two core sentences besides additional auxiliary sentences (e.g., an argument whose conclusion belongs to the debate's core while its premisses don't). There are, in total, 20 distinct arguments of this kind.[2] Whether such an argument passes the deductive-validity check or not depends on the correct truth values of its premisses and conclusion, including the core sentences' truth values. The following analysis is going to demonstrate: The higher the verisimilitude of the core position, the

[2]These combinatorial possibilities may be enumerated as follows. Assume q_1, q_2 were core sentences of a debate. The arguments we consider comprise these sentences or their negations. To count the arguments, we simply have to determine the different roles the core sentences may play, assuming that additional positions in the argument are filled with auxiliary assumptions. So there are eight $(4 \cdot 2)$ arguments that relate to exactly one of the two core sentences ($q_1, q_2, \neg q_1$, or $\neg q_2$ might either figure as the conclusion or as a premiss). In addition, there are four $(2 \cdot 2)$ arguments with two core sentences as premisses (q_1 or $\neg q_1$ as first premiss, q_2 or $\neg q_2$ as second one). Finally, there are eight $(4 \cdot 2)$ arguments with a core conclusion and an additional core sentence as premiss ($q_1, q_2, \neg q_1$, or $\neg q_2$ might figure as the conclusion; in each case, there are two choices left to pick the core premiss, e.g., q_1 or $\neg q_1$ if $\neg q_2$ is the conclusion).

Table 15.1 Deductive validity of arguments which tend to decrease the degree of justification of a core position which considers q_1 and q_2 true. Whether such an argument is deductively valid (in the semantic sense) depends on the actual truth values of q_1 and q_2 and is indicated in the table by "1" (valid) and "0" (invalid). The auxiliary sentences p_1 and p_2 are assumed to be correct

		Correct truth values of q_1 and q_2, respectively			
No.	Argument	$\frac{q_1 q_2}{FF}$	$\frac{q_1 q_2}{FT}$	$\frac{q_1 q_2}{TF}$	$\frac{q_1 q_2}{TT}$
1	$(p_1, p_2; \neg q_1)$	1	1	0	0
2	$(p_1, p_2; \neg q_2)$	1	0	1	0
3	$(p_1, q_1; \neg p_2)$	1	1	0	0
4	$(p_1, q_2; \neg p_2)$	1	0	1	0
5	$(p_1, q_1; \neg q_2)$	1	1	1	0
6	$(p_1, q_2; \neg q_1)$	1	1	1	0
7	$(q_1, q_2; \neg p_1)$	1	1	1	0
Total		7	5	5	0

more arguments which increase its degree of justification, and the less arguments which decrease its degree of justification, are deductively valid. As proponents may introduce but deductively valid arguments, this explains why core positions with a high verisimilitude tend to possess a high degree of justification, and vice versa.

Table 15.1 lists all arguments which contain at least one of the core sentences q_1, q_2 and which tend to decrease the degree of justification of a core position stating that q_1 and q_2 are true. Let us briefly verify this claim. Some of the arguments express the very same inferential relation. Thus, arguments 1 and 3, arguments 2 and 4, as well as arguments 5–7 are logically equivalent. So we just have to understand why 1, 2, and 7 decrease the degree of justification of a partial position which considers q_1 and q_2 true. As arguments 1 and 2 represent direct attacks against one of the core position's theses, these arguments typically reduce the degree of justification (see Sect. 2.2). Argument 7, however, amounts to a (potential) reductio of the core position, assuming the core beliefs as premisses, and therefore tends to decrease its the degree of justification.

The auxiliary sentences p_1 and p_2 are assumed to be correct (while the background knowledge remains empty), to the effect that the table only contains arguments with true auxiliary premisses and false auxiliary conclusion. Otherwise, the argument's deductive validity wouldn't depend on the truth values of q_1 and q_2. Table 15.1 indicates for each argument in which case—regarding the truth values of q_1 and q_2—it passes the deductive-validity check. If both core sentences are actually false, all arguments are deductively valid. If precisely one of the core sentences is false, 5 out of 7 arguments are deductively valid. And if both core sentences are actually correct, then no argument listed in the table is deductively valid. In sum, the closer q_1 and q_2 are to the truth, the less arguments which potentially decrease their degree of justification are deductively valid.

Table 15.2 Deductive validity of arguments which tend to increase the degree of justification of a core position which maintains q_1 and q_2. Whether such an argument is deductively valid (in the semantic sense) depends on the actual truth values of q_1 and q_2, and is indicated in the table by "1" (valid) and "0" (invalid). The auxiliary sentences p_1 and p_2 are assumed to be correct

		Correct truth values of q_1 and q_2, respectively			
No.	Argument	$\frac{q_1 q_2}{FF}$	$\frac{q_1 q_2}{FT}$	$\frac{q_1 q_2}{TF}$	$\frac{q_1 q_2}{TT}$
1	$(p_1, p_2; q_1)$	0	0	1	1
2	$(p_1, p_2; q_2)$	0	1	0	1
3	$(p_1, \neg q_1; \neg p_2)$	0	0	1	1
4	$(p_1, \neg q_2; \neg p_2)$	0	1	0	1
5	$(\neg q_1, \neg q_2; \neg p_1)$	0	1	1	1
6	$(p_1, \neg q_1; q_2)$	0	1	1	1
7	$(p_1, \neg q_2; q_1)$	0	1	1	1
8	$(\neg q_1, q_2; \neg p_1)$	1	0	1	1
9	$(q_1, \neg q_2; \neg p_1)$	1	1	0	1
10	$(p_1, q_1; q_2)$	1	1	0	1
11	$(p_1, q_2; q_1)$	1	0	1	1
12	$(p_1, \neg q_1; \neg q_2)$	1	0	1	1
13	$(p_1, \neg q_2; \neg q_1)$	1	1	0	1
Total		6	8	8	13

Table 15.2 presents a similar analysis with respect to all arguments which tend to increase the degree of justification of a core position that considers q_1 and q_2 true. Here, arguments 1 and 3; arguments 2 and 4; arguments 5–7; arguments 8, 11, and 12; as well as arguments 9, 10, and 13 express, respectively, the very same inferential relation between the corresponding three sentences and are thence logically equivalent. The arguments 1 and 2 directly support a core position's claim and consequently increase its degree of justification. The arguments 5, 8, and 9 represent (potential) reductio arguments directed against a rival alternative of the core position we consider. Argument 5, for example, decreases the degree of justification of the partial position according to which q_1 and q_2 are both false (see also argument 5 in Table 15.1), without rendering any complete position incoherent that extends our core position. It therefore increases the degree of justification of the core position which states that q_1 and q_2 are true.

Which of these arguments are deductively valid depends on the correct truth values of q_1 and q_2 (again, we assume that the auxiliary sentences p_1 and p_2 are both correct). According to Table 15.2, merely 6 out of 13 arguments are valid if both core sentences are false. If exactly one of the two core sentences is false, eight arguments are valid. And in case both q_1 and q_2 are correct, then all arguments are deductively valid. Thus, the closer q_1 and q_2 are to the truth, the more arguments which tend to increase the degree of justification of a core position containing q_1 and q_2 are valid and may thus be introduced into the debate.

So far, we have considered but arguments with one or two core sentences. But the analysis can easily be extended to arguments containing three core sentences. Such an argument renders exactly one truth-value assignment to these core sentences incoherent. It increases, *ceteris paribus*, the degree of justification of the partial position according to which the three core sentences are true, if it renders a rival position incoherent. It decreases the degree of justification (to 0) if the core position itself is rendered incoherent. Yet the latter cannot happen if the core position is fully correct.[3] Thus, proximity to truth decreases the number of arguments which reduce the degree of justification and increases, vice versa, the number of robustness-raising arguments.

In sum, the higher a core position's verisimilitude, the less deductively valid arguments may be introduced into the debate which decrease its degree of justification, and the more deductively valid arguments there are which increase its degree of justification. In a debate where newly introduced arguments are drawn randomly from the pool of potential arguments, the degree of justification of mainly correct core positions will therefore rise faster than the degree of justification of rather incorrect core positions.

Let us close this discussion with a caveat. The analysis carried out hitherto should be understood as a conceptual one. It provides merely a qualitative explanation of the main result that degree of justification correlates with verisimilitude. That is because the arguments enlisted in Tables 15.1 and 15.2 only *tend* to decrease or increase the degree of justification of the corresponding core position and may very well bring about the opposite in special situations. More specifically, only if an argument from those lists is independent of the other arguments which have been introduced into the debate before (i.e., precisely, if its premisses or their negations don't figure in other arguments), does it *necessarily* decrease or, respectively, increase the corresponding degree of justification. Yet if that is not the case, the effect of introducing the argument cannot be predicted with certainty independently of the dialectic context. However, the argument might still *tend* to alter the degree of justification in specific ways. More importantly, the assumption that new arguments are by and large independent seems to be not completely unrealistic given the low inferential densities ($D \leq 0.15$) we are considering here.

[3] We have noted this asymmetry before, in Sect. 13.3, where it helped to explain the truth-conduciveness of critical argumentation.

Chapter 16
The Veritistic Dynamics of Debates with Core Argumentation

Having studied the veritistic dynamics of random debates with *lexicographic* update mechanism in the previous chapter, we will consider, in this chapter, argumentation strategies that take the distinction between core and auxiliary sentences explicitly into account. More specifically, we modify, firstly, the highly truth-conducive *t-multiple convert* strategy (cf. Chap. 14) with a view to a debate's core sentences and reconsider, secondly, the argumentation strategy which instructs a proponent to maximize the degree of justification of her core position (see Chap. 9). One of our chief interests consists in investigating whether a core position's robustness remains an accurate indicator of truth when proponents employ the sophisticated argumentation strategies.

16.1 Setup

We study two ensembles with 1,000 debates each. Every debate contains six proponents. The sentence pool, S, comprises five core sentences, $C \subset S$, and 15 auxiliary sentences. The debate simulations terminate if all proponents have adopted the true position $\mathscr{T}$.

The first ensemble serves to investigate the modified *t-multiple convert* rule. Its debates are set up as follows:

Argumentation Mechanism: The proponents introduce, in successive order, new arguments in line with the following rule: Unless all proponents agree, the proponent i chooses at step t, randomly, one of her core beliefs—this makes up the conclusion c of the new argument ($c \in C$ and $\mathscr{P}_t^i(c) = \text{true}$). She identifies, subsequently, all pairs of sentences which, if taken as premisses, yield a valid argument, that is, an argument that leaves the true position coherent. From these, finally, a pair of sentences which most opponents adhere to is chosen; it constitutes the new argument's premisses. If, however, all proponents agree,

G. Betz, *Debate Dynamics: How Controversy Improves Our Beliefs*,
Synthese Library 357, DOI 10.1007/978-94-007-4599-5_16,

the newly introduced argument is constructed so as to render the consensus position incoherent, if possible. We call this argumentation strategy *t-multiple core convert*.

Discovery Mechanism: There is no background knowledge.
Update Mechanism: *Lexicographic closest coherent* (cf. Sect. 8.1).

In the second ensemble, proponents attempt to maximize the robustness of their current core position. The debates are set up as follows:

Argumentation Mechanism: The proponents put forward, in alternating sequence, one argument each. The new argument is valid, leaving the true position coherent, and maximizes—relative to all arguments the proponent could alternatively introduce—the robustness of the corresponding proponent's core position.[1] We shall call this strategy *t-robust argumentation*.

Discovery Mechanism: There is no background knowledge.
Update Mechanism: *Lexicographic closest coherent* (cf. Sect. 8.1).

16.2 Results

16.2.1 Core Truth-Conduciveness

The left-hand panel of Fig. 16.1 displays the mean verisimilitude evolutions of the proponents' core positions for this chapter's ensembles as well as for the ensemble studied in the previous chapter. The *t-multiple core convert* rule gives rise to a substantial verisimilitude growth at relatively low densities. Accordingly, at a density of $D = 0.5$, the mean verisimilitude of proponent core positions has increased by 35 percentage points to a level of 0.85. At the same density, the verisimilitude of core positions in the ensembles with *t-robust argumentation* and *t-random argumentation* amounts to merely 0.55, in contrast. Surprisingly, *t-robust argumentation* performs even worse than *t-random argumentation* in terms of truth-conduciveness, in particular at high densities.

In terms of the number of fully correct proponent positions, the difference between the strategies is no less stunning. As the right-hand plot in Fig. 16.1 shows, proponents start to acquire the fully correct core position even at low densities in the ensemble with *t-multiple core convert*. At a density of $D = 0.5$, more than 4 (out of 6) proponents have reached the truth. This compares, at the same density, with less than 1 proponent in the ensembles with *t-robust argumentation* and *t-random argumentation*. Yet, and this marks a difference to the verisimilitude evolutions in the left-hand plot of Fig. 16.1, *t-robust argumentation* is not consistently outperformed by *t-random argumentation*: At densities lower than 0.8, slightly

[1] Technically, the number of different arguments whose effect on the core position's degree of justification we calculate at each time step is limited to 100 because of computational constraints.

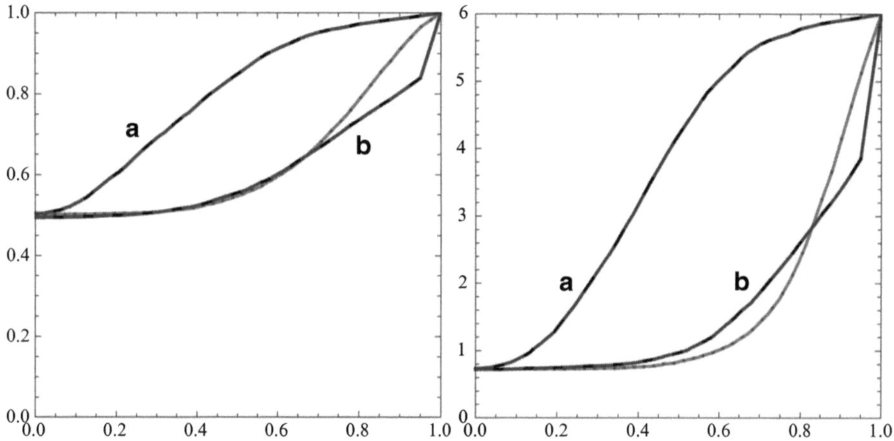

Fig. 16.1 Ensemble-wide mean verisimilitude of proponent core positions (*left*) and ensemble-wide mean number of fully correct proponent core positions (*right*) as functions of inferential density. Average normalized verisimilitude and number of fully correct proponent cores are plotted for this chapter's ensembles with *t-multiple core convert* (*a*) and *t-robust argumentation* (*b*), as well as for the ensemble, presented in Chap. 15, with *t-random argumentation* (*gray curve*)

more proponents who maximize the robustness of their position have found the truth as compared to proponents who discover arguments randomly. At higher densities, though, debates with *t-random argumentation* tend to contain more proponents with a fully correct core position.

Let us next consider how the robustness of the core positions affects the veritistic dynamics. Like in previous analyses, we evaluate the degree of justification of the proponent core positions at a density of $D = 0.15$. We distinguish proponents whose core position is very robust (upper quartile) and very unrobust (lower quartile). Figure 16.2 demonstrates that, at $D = 0.15$, robust core positions are, on average and in both ensembles, closer to the truth. Moreover, proponents who hold a robust core position at $D = 0.15$ typically take off from an initial position with above-average verisimilitude, too, and the core positions they adopt at higher densities, at least up to $D = 0.5$, display closer proximity to the truth than the positions adopted by their counterparts (who maintain an unrobust position at $D = 0.15$). At high densities, however, having held a very robust core position at an early stage of the debate gradually ceases to be advantageous in terms of verisimilitude.

Figure 16.3 demonstrates that the robustness of a proponent core position affects the likelihood that the proponent adopts, or will adopt, a fully correct core position. More precisely, the number of proponents who maintain a robust position at $D = 0.15$ and hold a completely true core position at the density D' is—for all densities $0 \leq D' \leq 1$—at least as great as the number of proponents with an unrobust position at $D = 0.15$ and a fully correct core position at D'. At high densities, however, as all proponents acquire a fully correct core position, the difference between proponents with a robust respectively unrobust position at $D = 0.15$ shrinks.

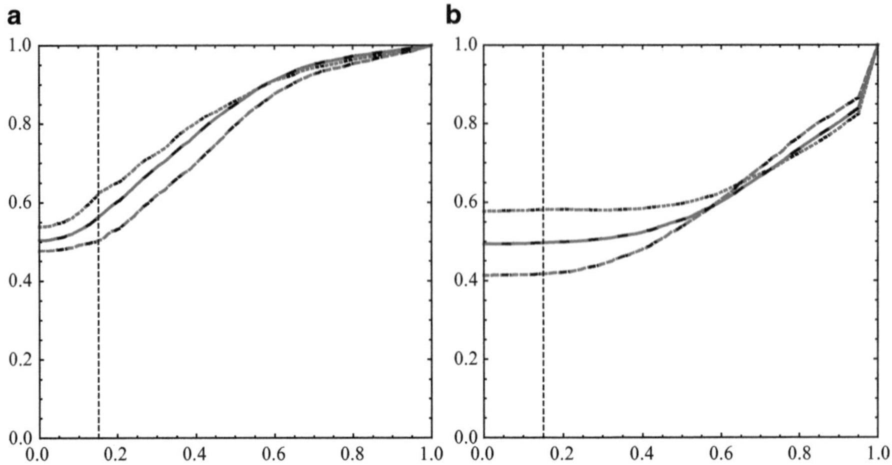

Fig. 16.2 Ensemble-wide mean verisimilitude of proponent core positions as a function of inferential density, plotted for this chapter's ensembles with *t-multiple core convert* (**a**) and *t-robust argumentation* (**b**). The plots display ensemble-wide means as averaged over all proponents (*solid curves*), proponents with a very robust core position at $D = 0.15$ (*dotted curves*), and proponents who hold a core position with very low robustness at $D = 0.15$ (*dashed curves*). More specifically, a partial core position with high (low) robustness possesses a degree of justification which falls in the upper (lower) quartile of all robustness scores at the corresponding density in the ensemble

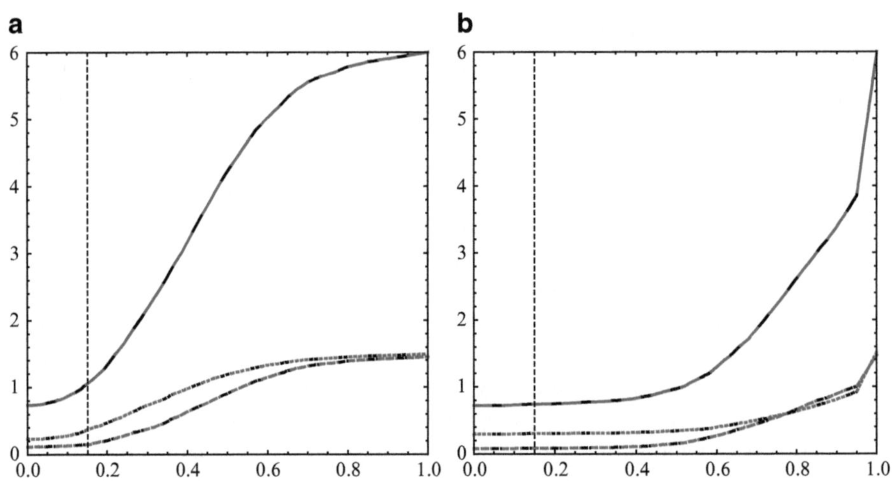

Fig. 16.3 Ensemble-wide mean number of fully correct proponent core positions as a function of inferential density, plotted for this chapter's ensembles with *t-multiple core convert* (**a**) and *t-robust argumentation* (**b**) (see Fig. 16.2 for further information)

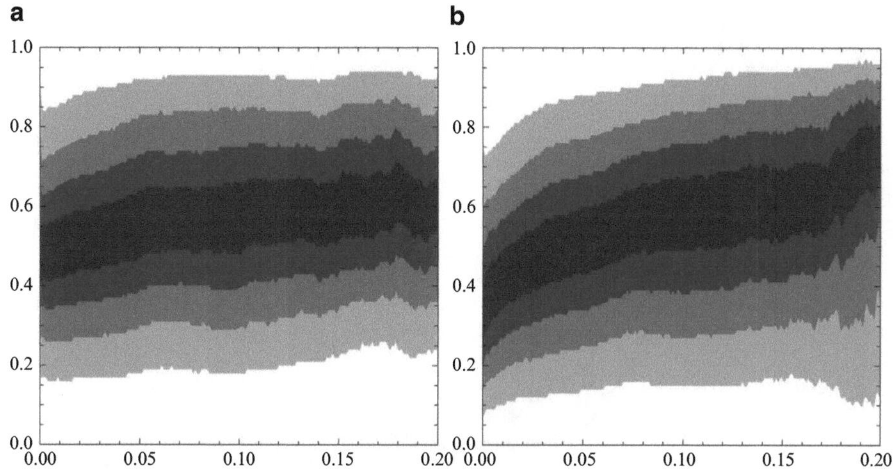

Fig. 16.4 Ensemble-wide mean verisimilitude of proponent core positions as a function of their robustness at a density of $D = 0.15$, plotted for this chapter's ensembles with *t-multiple core convert* (**a**) and *t-robust argumentation* (**b**). In order to increase its accuracy, these plots are based on an ensemble of at least 4,000 instead of 1,000 debate simulations (compare Fig. 15.3 for further information)

16.2.2 Robustness of Proponent Core Positions and Verisimilitude

We have found, in the previous chapter, a positive correlation between a proponent core position's degree of justification and its verisimilitude. The results presented so far, in particular Figs. 16.2 and 16.3, suggest that a similar relationship also holds in this chapter's ensembles. Is degree of justification, in these ensembles, maybe even more accurate an indicator of truth than in the ensemble with *t-random argumentation*? The brief answer to this question, provided by Fig. 16.4, is no. And this is true in spite of the fact that both *t-multiple core convert* and *t-robust argumentation* give rise to much more extreme robustness scores than *t-random argumentation*, as shown in the histogram Fig. 16.5.[2] In the ensemble with *t-multiple core convert* (Fig. 16.4a), the relationship between degree of justification and verisimilitude is not closer, but seriously less pronounced than in the ensemble with *t-random argumentation* (compare Fig. 15.3). In the former case, merely 46% of the very robust core positions (i.e., core positions with a degree of justification greater than 0.19, which represent ca. 2% of all positions) display a verisimilitude greater than 0.8, as compared to 60% of the very robust core positions (degree of justification greater than 0.09, accounting for roughly 1% of all positions) in the

[2]That is also the reason why the plots in Fig. 16.4 range over a robustness interval from 0.0 to 0.2, whereas Fig. 15.3 covers merely the robustness interval $[0; 0.1]$.

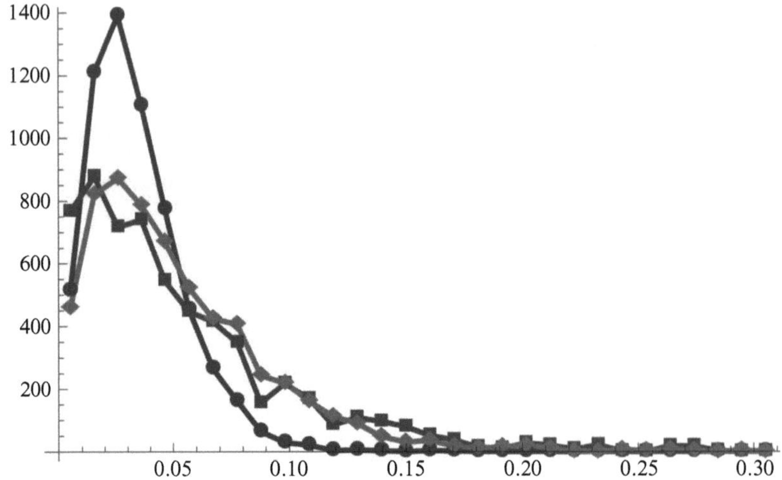

Fig. 16.5 Absolute frequencies of proponent core positions with corresponding robustness values at $D = 0.15$. Frequency distributions are plotted for this chapter's ensembles with *t-multiple core convert* (*squares*) and *t-robust argumentation* (*diamonds*), as well as for the ensemble, presented in Chap. 15, with *t-random argumentation* (*circles*)

ensemble with *t-random argumentation*. Moreover, core positions with an extremely low degree of justification tend to be rather false in the ensemble with *t-random argumentation*, yet possess an average (and hence uninformative) verisimilitude in the ensemble with *t-multiple core convert*.

Unlike in the ensemble with *t-multiple core convert*, the degree of justification is, with *t-robust argumentation*, at least as telling an indicator of truth as with *t-random argumentation* (compare Figs. 16.4b and 15.3). In particular, 66% of the highly robust core positions (i.e., positions with a degree of justification greater than 0.19, accounting for 1% of all core positions) display a verisimilitude greater than 0.8 in the ensemble with *t-robust argumentation*, as a statistical analysis of the ensemble reveals. Similarly, a core position with a very low degree of justification is likely to be rather false—in agreement with the findings regarding random debates, where very low robustness accurately indicates very low verisimilitude (see Fig. 15.3). But although robustness remains an important indicator of truth, *t-robust argumentation* doesn't succeed in fostering the accuracy of this indicator substantially. It differs, in this respect, from *t-multiple undercut*, which increased the accuracy of both consensus and stability as indicators of truth. In addition, we may note a major difference between the veritistic and consensus dynamics, as well, since *robust argumentation* does in fact increase the accuracy of a position's degree of justification as an indicator of proximity to the final consensus (see Fig. 9.8).

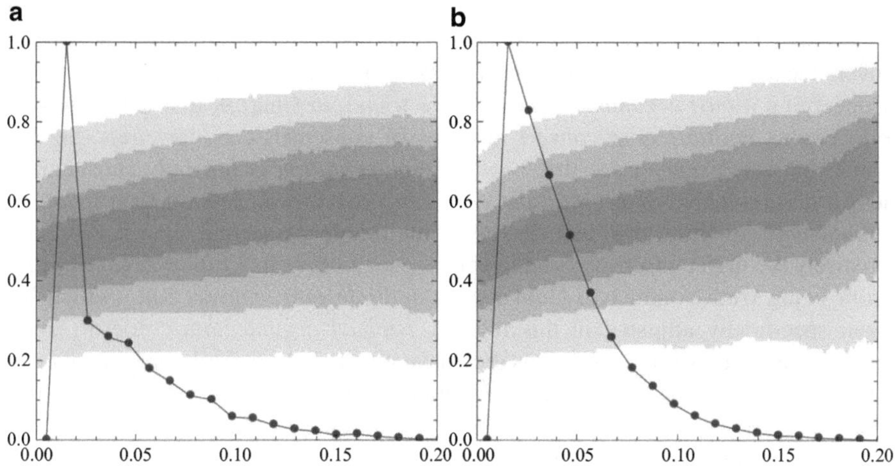

Fig. 16.6 Ensemble-wide mean verisimilitude of all core positions as a function of their robustness at a density of $D = 0.15$, plotted for this chapter's ensembles with *t-multiple core convert* (**a**) and *t-robust argumentation* (**b**). In order to increase its accuracy, these plots are based on an ensemble of at least 4,000 instead of 1,000 debate simulations (compare Fig. 15.4 for further information)

16.2.3 General Correlation Between Degree of Justification and Verisimilitude

Figure 16.6 depicts the relationship between verisimilitude and degree of justification, taking account of all core positions in the debates, and not merely of those which are held by some proponent, as in Fig. 15.4. We observe, at most, a very weak positive association in the ensemble with *t-multiple core convert* (a), yet can identify a solid positive correlation in the ensemble with *t-robust argumentation* (b). Hence, the significance of robustness as an indicator of truthlikeness—namely, the truthlikeness both of all core positions and of the proponent core positions—depends on the specific argumentation strategies employed by the proponents.

16.3 Discussion

The results presented in the previous section raise two points which deserve further discussion. The first item relates to the poor performance of the *t-robust argumentation* rule in terms of increasing average core verisimilitude in the course of a debate. The second point pertains to the way the core argumentation strategies, studied in this chapter, alter the accuracy of robustness as an indicator of truth.

The rapid and substantial verisimilitude increase of core positions with *t-multiple core convert* observed in the previous section comes hardly as a surprise, given the outstanding truth-conduciveness of *t-multiple convert* (cf. Chap. 14). Concerning *robust argumentation*, we have found, in Chap. 9, that it is almost as effective as *multiple core convert* in terms of consensus-conduciveness. So it is somewhat surprising to see that *t-robust argumentation* is not only substantially less truth-conducive than *t-multiple core convert* but even less so than *t-random argumentation*. Still, this fact becomes intelligible at second glance. For it is precisely the high consensus-conduciveness which prevents *t-robust argumentation* from being truth-conducive. Unlike the *t-multiple core convert* rule (which has been specifically adjusted in this regard), *t-robust argumentation* doesn't invite proponents to question a consensus they have reached. As a result, proponents who implement the *t-robust argumentation* strategy run the risk of being caught in a spurious consensus. Thus, it is the lack of (self-)critical elements which explains the poor truth-conduciveness of a homogeneous debate with *t-robust argumentation*.

As the main finding of this chapter, we have established that the accuracy of robustness as an indicator of truth is sensitive to the argumentation strategies employed by the proponents. Whereas, with *t-random argumentation*, degree of justification is positively correlated with verisimilitude, this is hardly the case if proponents follow the *t-multiple core convert* rule. Finally, *t-robust argumentation* doesn't seem to affect the *overall* accuracy of this veritistic indicator (as compared to *t-random argumentation*), yet alters it in a particular way: With *t-robust argumentation*, high (low) degree of justification becomes more (less) accurate an indicator of truth.

Let us briefly recall the explanation for why robustness correlates with verisimilitude in case arguments are introduced randomly (cf. Sect. 15.3). As we have seen in the previous chapter, the higher the verisimilitude of a partial position at a given state of the debate, the greater (smaller) the proportion of potential—i.e., deductively valid—arguments which would increase (decrease) the partial position's robustness once introduced. Therefore, if arguments are drawn randomly from the pool of potential arguments, the robustness of rather true core positions tends to be increased more frequently, and decreased less frequently, than the robustness of rather false core positions. Degree of justification being an indicator of truth thus depends sensitively, at least in the case just considered, on the random selection of new arguments, that is, the specific argumentation mechanism employed. Now, in debates with *t-robust argumentation*, arguments aren't selected randomly from the set of all potential arguments, but are rather chosen, by a proponent, so as to maximize her corresponding core position's degree of justification. So the correlation between robustness and verisimilitude, which pertains nonetheless, has to be explained differently. Apparently, proponents can much more successfully or effectively increase the robustness of their core position in case the latter is close to the truth. Our previous analysis, summarized above, suggests the following explanation for this fact: The higher the verisimilitude of a core position a proponent holds, the more potential (deductively valid) arguments which increase her position's robustness are at her disposal. Holding a predominantly correct core position thence represents a strategic advantage when it comes to maximizing one's position's robustness. The most robust core positions are those which are held by proponents who succeeded best in increasing their position's degree of justification, and these in turn are proponents with a primarily correct core position. That is why a high degree of justification indicates accurately high verisimilitude with *t-robust argumentation*. We have observed, moreover, that a low degree of justification is less accurate an indicator of truth with *t-robust argumentation* (compared to *t-random argumentation*). I suggest to explain this observation along the following lines. In a debate with *t-robust argumentation*, a proponent i holds a core position with a low degree of justification because at least one of the opponents j, $j \neq i$, has very successfully increased her core position's degree of justification, $\text{DOJ}(\mathcal{P}^j)$, thereby decreasing the robustness of the proponent's position, $\text{DOJ}(\mathcal{P}^i)$. Accordingly, low robustness results primarily from facing a successful opponent and merely secondarily from lacking effective means for increasing one's position's robustness. Whereas the latter depends on the verisimilitude of one's core position, the former doesn't—at least not to the same degree.[3]

In addition, we have to understand why robustness ceases to be a significant indicator of truth in debates with *t-multiple core convert*. First of all, we shall try to understand, generally, why some core positions become extremely robust or un-robust, if the proponents follow the *t-multiple core convert* rule. According to the *t-multiple core convert* strategy, a proponent introduces an argument which supports her core position and which builds on premisses that are agreed upon by as many opponents as possible. As a consequence, if many proponents hold initially (and hence coincidentally) one and the same, or a similar, core position, this very core position will receive vast argumentative support; actually, most of the arguments introduced will back up the individual claims of this very core position and consequently increase its degree of justification. In the same time, the degree of justification of the complementary core position will be radically reduced. Now, crucially, whether some core position is continously supported or incessantly attacked by the proponents, who apply the *t-multiple core convert rule* faithfully, is entirely determined by the randomly chosen initial positions of the proponents as well as their early agreement evolutions. The proximity to the truth, that is, has no bearing on this process whatsoever. Moreover, truth doesn't impose substantial constraints on the availability of potential arguments which support some consensus core position, because proponents can always construct deductively valid arguments (with actually false premisses) which increase the degree of justification of a given core position. This is why, with *t-multiple core convert*, proponent core positions possess extreme degrees of justification by and large independently of their verisimilitude.

The explanations advanced so far apply, strictly speaking, to core positions which are held by some proponent only. Yet we have found that the strength of

[3]Arguably, the success of the opponent hinges on the opponent's core position's verisimilitude. But the verisimilitudes of the proponent and the opponent position are not necessarily inversely related.

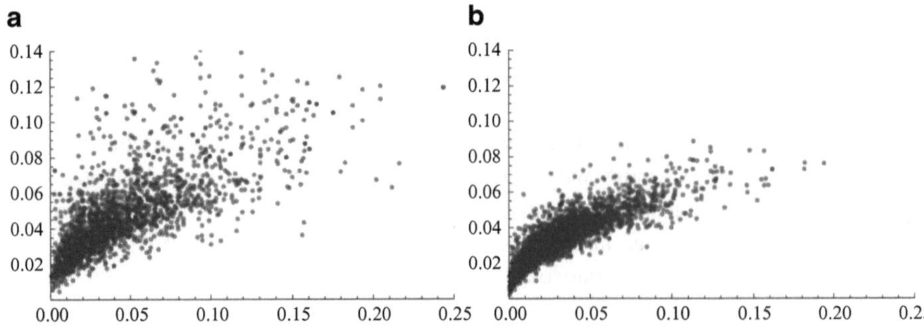

Fig. 16.7 Correlation between a core position's robustness and its neighbors' robustness, plotted for this chapter's ensembles with *t-multiple core convert* (**a**) and *t-robust argumentation* (**b**). For each coherent core position $\mathscr{P}$ at $D = 0.15$, the core's robustness (x-axis) is plotted against the average robustness of its adjacent core positions (y-axis)—whereas an adjacent core position disagrees with $\mathscr{P}$ with respect to exactly one core sentence. The plot considers all coherent core positions from a random sample of 100 debates drawn from the corresponding ensemble

the general correlation between robustness and verisimilitude—regarding all core positions, no matter whether they are maintained by some proponent or not—is sensitive to the argumentation strategy employed, too. To explain this fact, we may pick up a finding of Chap. 8, where we observed a neat correlation between a core position's robustness and the robustness of its neighbors (cf. Fig. 8.7). Because a similar correlation holds in this chapters' ensembles as well (see Fig. 16.7) and because the verisimilitude values of adjacent core positions are obviously correlated, a strong (weak) correlation between the proponent core positions' robustness and their verisimilitude spills over into a general strong (weak) association between robustness and verisimilitude.

Symbols

τ	dialectical structure
T	set of arguments in τ
A	attack relation on τ
U	support relation on τ
S_τ	sentences that figure in arguments in τ
S	pool of sentences
n	number of sentence pairs (sentence plus its negation) in such a pool
$\mathscr{P}, \mathscr{Q}, \ldots$	individual proponent positions (boolean vectors)
$[p_1, \ldots, p_n]$	position that assigns $p_1, \ldots, p_n$ the truth value *true*
$\mathscr{T}$	the true complete position, the truth
$\mathscr{B}$	partial position considered as basic background knowledge
$\beta, \beta_{\text{eff}}$	proportion of basic/effective background knowledge
$\mathbf{A}, \mathbf{B}, \ldots$	sets of positions
Γ_τ	set of coherent positions on τ (relative to some S)
$\Gamma_\tau(\mathscr{P}, \mathscr{Q}, \ldots)$	set of coh. pos. which extend $\mathscr{P}$ and $\mathscr{Q}$ and ...
σ_τ	number of coherent positions on τ (relative to some S)
$\sigma_\tau(\mathscr{P}, \mathscr{Q}, \ldots)$	number of coh. pos. which extend $\mathscr{P}$ and $\mathscr{Q}$ and ...
$p, p_1, p_2, \ldots, q, q_1, \ldots$	sentences, typically premisses
$c, c_1, c_2, \ldots$	sentences, typically conclusions
$\text{HD}(\mathscr{P}, \mathscr{Q})$	Hamming distance between two positions
$\Delta(\mathscr{P}, \mathscr{Q})$	normalized distance between two positions
$D(\tau)$	inferential density of τ
$\text{NCC}(\mathscr{P}, \mathbf{A})$	normalized closeness centrality of $\mathscr{P}$ relative to $\mathbf{A}$
t	step in a debate evolution
$\tau(t), \tau_t, \mathscr{P}(t), \mathscr{P}_t, \ldots$	respective values at step t
$\text{DOJ}_\tau(\mathscr{P})$	degree of justification of $\mathscr{P}$ in τ
$\text{DOJ}_\tau(\mathscr{P} \mid \mathscr{Q})$	conditional degree of justification of $\mathscr{P}$ given $\mathscr{Q}$

G. Betz, *Debate Dynamics: How Controversy Improves Our Beliefs*,
Synthese Library 357, DOI 10.1007/978-94-007-4599-5,

References

Alchourron, Carlos E., Peter Gärdenfors, and David Makinson. 1985. On the logic of theory change – partial meet contraction and revision functions. *Journal of Symbolic Logic* 50(2): 510–530.

Arrow, Kenneth J. 1963. *Social choice and individual values*. New York: Wiley.

Bench-Capon, Trevor J.M., and Paul E. Dunne. 2007. Argumentation in artificial intelligence. *Artificial Intelligence* 171(10–15): 619–641.

Besnard, Philippe, and Anthony Hunter. 2008. *Elements of argumentation*. Cambridge: MIT.

Betz, Gregor. 2008. Evaluating dialectical structures with Bayesian methods. *Synthese* 163(1): 25–44.

Betz, Gregor. 2009. Evaluating dialectical structures. *Journal of Philosophical Logic* 38: 283–312.

Betz, Gregor. 2010. *Theorie dialektischer Strukturen*. Frankfurt am Main: Klostermann.

Betz, Gregor. 2011a. *René Descartes: Meditationen über die Grundlagen der Philosophie. Ein systematischer Kommentar*. Stuttgart: Reclam.

Betz, Gregor. 2011b. On degrees of justification. *Erkenntnis*. forthcoming. dx.doi.org/10.1007/s10670-011-9314-y

Betz, Gregor. 2011c. Revamping hypothetico-deductivism: A dialectic account of confirmation. *Manuscript*. http://philsci-archive.pitt.edu/id/eprint/8822.

Betz, Gregor. 2011d. Justifying inference to the best explanation as a practical meta-syllogism on dialectical structures. *Manuscript*. http://philsci-archive.pitt.edu/id/eprint/8821.

Bondarenko, Andrei, Phan Minh Dung, Robert A. Kowalski, and Francesca Toni. 1997. An abstract, argumentation-theoretic approach to default reasoning. *Artificial Intelligence* 93(1–2): 63–101.

Carnap, Rudolf. 1950. *Logical foundations of probability*. Chicago: University of Chicago Press.

Carnap, Rudolf. 1956. *Meaning and necessity: A study in semantics and modal logic*, Chapter empiricism, semantics and ontology, 205–221. Chicago: Chicago University Press.

Cayrol, Claudette, and Marie-Christine Lagasquie-Schiex. 2005. On the acceptability of arguments in bipolar argumentation frameworks. In *Symbolic and quantitative approaches to reasoning with uncertainty*, ed. Lluís Godo. *Proceedings of the 8th European conference, ECSQARU 2005, Barcelona, Spain, July 6–8, 2005*, 378–389. Berlin/Heidelberg: Springer.

Cevolani, Gustavo, Vincenzo Crupi, and Roberto Festa. 2011. Verisimilitude and belief change for conjunctive theories. *Erkenntnis* 75(2): 183–202.

Chesñevar, Carlos I., Ana G. Maguitman, and Ronald P. Loui. 2000. Logical models of argument. *ACM Computing Surveys* 32(4): 337–383.

de Lavalette, Gerard R. Renardel, and Sjoerd D. Zwart. 2011. Belief revision and verisimilitude based on preference and truth orderings. *Erkenntnis* 75(2): 237–254

G. Betz, *Debate Dynamics: How Controversy Improves Our Beliefs*,
Synthese Library 357, DOI 10.1007/978-94-007-4599-5,

Descartes, René. 1984. Rules for the direction of the mind. In *The philosophical writings of Descartes*, vol. 1, ed. John Cottingham, Robert Stoothoff, and Dugald Murdoch, 9–78. Cambridge/New York: Cambridge University Press.

Douven, Igor, and Christoph Kelb. 2011. Truth approximation, social epistemology, and opinion dynamics. *Erkenntnis* 75(2): 271–283.

Dung, Phan Minh. 1995. On the acceptability of arguments and its fundamental role in nonmonotonic reasoning, logic programming and n-person games. *Artificial Intelligence* 77(2): 321–357.

Fagin, Ronald, Joseph Y. Halpern, Yoram Moses, and Moshe Vardi. 1995. *Reasoning about knowledge*. Cambridge: MIT.

Gärdenfors, Peter. 1988. *Knowledge in flux: Modeling the dynamics of epistemic states*. Cambridge: MIT.

Goldman, Alvin I. 1999. *Knowledge in a social world*. Oxford/New York, Oxford University Press.

Hansson, Sven O. 1999. *A textbook of belief dynamics: Theory change and database updating*, Applied logic series, vol. 11. Dordrecht: Kluwer.

Hansson, Sven O. 2009. Logic of belief revision. In *The Stanford Encyclopedia of Philosophy*, ed. Edward N. Zalta. Spring 2009 edition. http://plato.stanford.edu/archives/spr2009/entries/logic-belief-revision/.

Hegselmann, Rainer. 2004. Opinion dynamics – insights by radically simplifying models. In *Laws and models in science*, ed. Donald Gillies, 19–46. London: King's College Publications.

Hegselmann, Rainer, and Ulrich Krause. 2002. Opinion dynamics and bounded confidence: Models, analysis and simulation. *Journal of Artificial Societies and Social Simulation* 5(3). http://jasss.soc.surrey.ac.uk/5/3/2.html.

Hegselmann, Rainer, and Ulrich Krause. 2006. Truth and cognitive division of labour: First steps towards a computer aided social epistemology. *Journal of Artificial Societies and Social Simulation*, 9(3). http://jasss.soc.surrey.ac.uk/9/3/10.html.

Hendricks, Vincent F. 2006. *Mainstream and formal epistemology*. Cambridge/New York: Cambridge University Press.

Hintikka, Jaakko. 1962. *Knowledge and belief: An introduction to the logic of the two notions*. Ithaca: Cornell University Press.

Kitcher, Philip. 2001. *Science, truth, and democracy*. Oxford/New York: Oxford University Press.

Kölbl-Ebert, Martina. 2003. From volcano to impact crater: A history of the impact hypothesis at Ries Crater and Steinheim Basin from 1900 to 1970. *Neues Jahrbuch für Geologie und Paläontologie-Monatshefte* 2003(10): 591–602.

Kuhn, Thomas S. 1962. *The structure of scientific revolutions*. Chicago: University of Chicago Press.

Kuipers, Theo, and Gerhard Schurz. 2011. Introduction and overview. *Erkenntnis* 75(2): 151–163.

Lehrer, Keith, and Carl Wagner. 1981. *Rational consensus in science and society*, Philosophical studies series in philosophy, vol. 24. Dordrecht: Reidel.

Levi, Isaac. 1991. *The fixation of belief and its undoing: Changing beliefs through inquiry*. Cambridge/New York: Cambridge University Press.

List, Christian, and Philip Pettit. 2002. Aggregating sets of judgments: An impossibility result. *Economics and Philosophy* 18(1): 89–110.

List, Christian, and Philip Pettit. 2004. Aggregating sets of judgments: Two impossibility results compared. *Synthese* 140(1–2): 207–235.

List, Christian, and Ben Polak. 2010. Introduction to judgment aggregation. *Journal of Economic Theory* 145(2): 441–466.

List, Christian, and Clemens Puppe. 2009. Judgment aggregation: A survey. In *The handbook of rational and social choice: An overview of new foundations and applications*, ed. Paul Anand, Prasanta K. Pattanaik, and Clemens Puppe, 457–482. Oxford/New York: Oxford University Press.

Mill, John S. 2009. *On Liberty / Über die Freiheit*. Stuttgart: Reclam.

Niiniluoto, Ilkka. 1998. Verisimilitude: The third period. *British Journal for the Philosophy of Science* 49: 1–29.

Niiniluoto, Ilkka. 2011. Revising beliefs towards the truth. *Erkenntnis* 75(2): 165–181.

Oddie, Graham. 2008. Truthlikeness. In *The Stanford Encyclopedia of Philosophy*, ed. Edward N. Zalta. Fall 2008 edition. http://plato.stanford.edu/archives/fall2008/entries/truthlikeness/.

Popper, Karl. 1963. Truth, rationality, and the growth of scientific knowledge. In *Conjectures and refutations*, 291–338. London: Routledge and Kegan Paul.

Prakken, Henry, and Gerard Vreeswijk. 2001. Logics for defeasible argumentation. In *Handbook of philosophical logic*, vol. 4, ed. Dov M. Gabbay, and Franz Guenthner, 219–318. Dordrecht: Kluwer.

Putnam, Hilary. 1981. *Reason, truth and history*. Cambridge: Cambridge University Press.

Reiter, Raymond. 1980. A logic for default reasoning. *Artificial Intelligence* 13(1–2): 81–132.

Riegler, Alexander, and Igor Douven. 2009. Extending the Hegselmann-Krause model III: From single beliefs to complex belief states. *Episteme* 6(2): 145–163.

Schleichert, Hubert. 1998. *Wie man mit Fundamentalisten diskutiert, ohne den Verstand zu verlieren. Anleitung zum subversiven Denken*. München: C.H. Beck.

Shoemaker, Eugene M., and Edward C.T. Chao. 1961. New evidence for impact origin of Ries Basin, Bavaria, Germany. *Journal of Geophysical Research* 66(10): 3371–3378.

Simon, Herbert A. 1982. *Models of bounded rationality*. Cambridge: MIT.

Thagard, Paul. 1992. *Conceptual revolutions*. Princeton: Princeton University Press.

van Ditmarsch, Hans, Wiebe van der Hoek, and Barteld P. Kooi. 2007. *Dynamic epistemic logic*, Synthese library. Dordrecht: Springer.

von Engelhardt, Wolf. 1982. Hypotheses on the origin of the Ries Basin, Germany, from 1792 to 1960. *Geologische Rundschau* 71(2): 475–486.

Zamora Bonilla, Jesús P. 1992. Truthlikeness without truth: A methodological approach. *Synthese* 93(3): 343–372.

Zamora Bonilla, Jesús P. 2000. Truthlikeness, rationality and scientific method. *Synthese* 122(3): 321–335.

Index

G. Betz, *Debate Dynamics: How Controversy Improves Our Beliefs*, Synthese Library 357, DOI 10.1007/978-94-007-4599-5,

MIX
Papier aus verantwortungsvollen Quellen
Paper from responsible sources
FSC® C105338

If you have any concerns about our products,
you can contact us on
ProductSafety@springernature.com

In case Publisher is established outside the EU,
the EU authorized representative is:
Springer Nature Customer Service Center GmbH
Europaplatz 3, 69115 Heidelberg, Germany

Printed by Libri Plureos GmbH
in Hamburg, Germany